Concepts of Mathematics and Physics

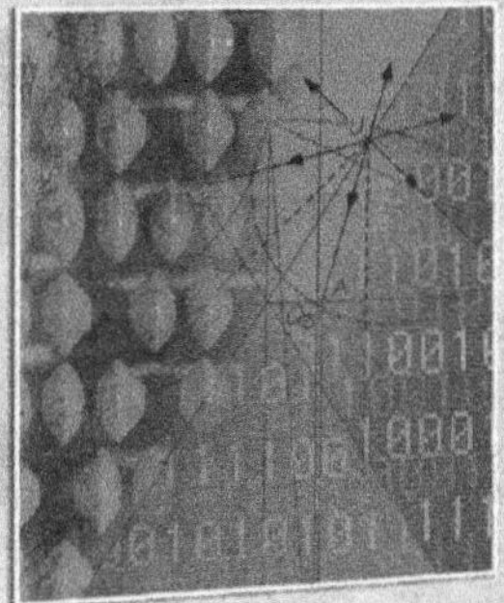

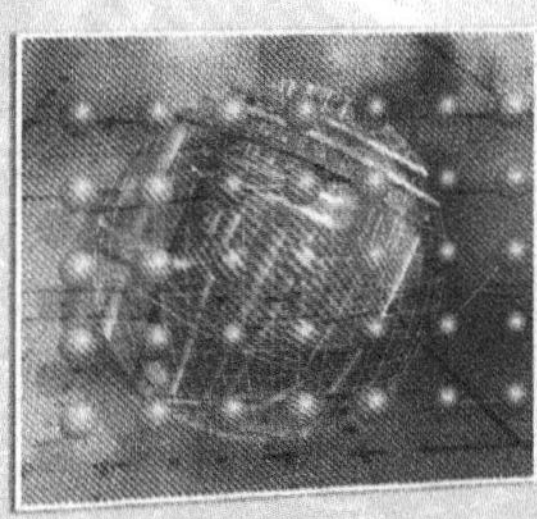

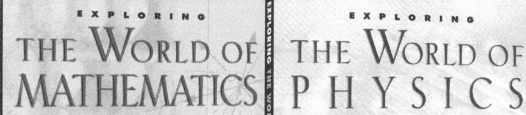

LESSON PLANNER

 Weekly Lesson Schedule

 Student Worksheets

 Quizzes & Test

 Answer Key

7th – 9th grade

1 Year Science

1/2 Credit

First printing: March 2013
Third printing: August 2015

Master Books®, P.O. Box 726, Green Forest, AR 72638

Master Books® is a division of the New Leaf Publishing Group, Inc.

ISBN: 978-0-89051-734-5

Unless otherwise noted, Scripture quotations are from the New King James Version of the Bible.

Printed in the United States of America

Please visit our website for other great titles:
www.masterbooks.com

For information regarding author interviews,
please contact the publicity department at (870) 438-5288

Since 1975, Master Books has been providing educational resources based on a biblical worldview to students of all ages. At the heart of these resources is our firm belief in a literal six-day creation, a young earth, the global Flood as revealed in Genesis 1–11, and other vital evidence to help build a critical foundation of scriptural authority for everyone. By equipping students with biblical truths and their key connection to the world of science and history, it is our hope they will be able to defend their faith in a skeptical, fallen world.

If the foundations are destroyed, what can the righteous do?
Psalm 11:3 NKJV

As the largest publisher of creation science materials in the world, Master Books is honored to partner with our authors and educators, including:

Ken Ham of Answers in Genesis

Dr. John Morris and Dr. Jason Lisle of the Institute for Creation Research

Dr. Donald DeYoung and Michael Oard of the Creation Research Society

Dr. James Stobaugh, John Hudson Tiner, Rick and Marilyn Boyer, Dr. Tom DeRosa, Todd Friel, Israel Wayne, and so many more!

Whether a pre-school learner or a scholar seeking an advanced degree, we offer a wonderful selection of award-winning resources for all ages and educational levels.

But sanctify the Lord God in your hearts, and always be ready to give a defense to everyone who asks you a reason for the hope that is in you, with meekness and fear.
1 Peter 3:15 NKJV

Permission to Copy

Contents

Lessons for a 36-week course!

Overview: This *Concepts of Mathematics and Physics PLP* contains materials for use with *Exploring the World of Mathematics* and *Exploring the World of Physics* in the Exploring series. Materials are organized by each book in the following sections:

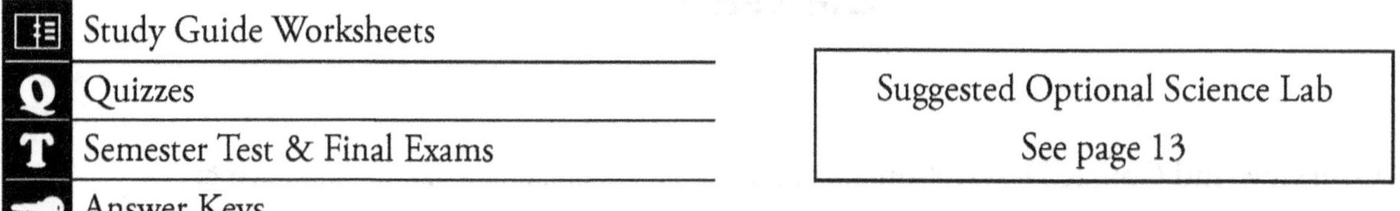

	Study Guide Worksheets
Q	Quizzes
T	Semester Test & Final Exams
	Answer Keys

Suggested Optional Science Lab
See page 13

Features: Each suggested weekly schedule has two easy-to-manage lessons which combine reading, worksheets, and vocabulary-building opportunities. Worksheets, quizzes, and tests are perforated and three-hole punched – materials are easy to tear out, hand out, grade, and store. As always, you are encouraged to adjust the schedule and materials as you need to in order to best work within your educational program.

Workflow: Students will read the pages in their book and then complete each section of the PLP. They should be encouraged to complete as many of the activities and projects as possible as well. Tests are given at regular intervals with space to record each grade. If used with younger students, they may be given the option of only choosing activities or projects of interest to them and taking open book tests.

Lesson Scheduling: Space is given for assignment dates. There is flexibility in scheduling. For example, the parent may opt for a M–W schedule rather than a M, W, F schedule. Each week listed has five days but due to vacations the school work week may not be M–F. Adapt the days to your school schedule. As the student completes each assignment, he/she should put an "X" in the box.

	Approximately 30 to 45 minutes per lesson, two days a week
	Includes answer keys for worksheets, quizzes, and semester exams
	Worksheet for each chapter.
	Quizzes are included to help reinforce learning and provide assessment opportunities; optional semester exams included.
	Designed for grades 7 to 9 in a one-year course to earn 1/2 science credit

Course includes books from authors with solid, biblical worldviews:

John Hudson Tiner — *Exploring the World of Mathematics, Exploring the World of Physics*

John Hudson Tiner received five National Science Foundation teaching fellowships during his 12 years as a teacher of mathematics and science that allowed him to study graduate chemistry, astronomy, and mathematics. He also worked as a mathematician and cartographer for the Defense Mapping Agency, Aerospace Center in St. Louis, MO.

Tiner has received numerous honors for his writing, including the Missouri Writer's Guild award for best juvenile book for *Exploring the World of Chemistry*. He and his wife, Jeanene, live in Missouri.

Concepts of Mathematics and Physics

Course Description

This is the suggested course sequence that allows one core area of science to be studied per semester. You can change the sequence of the semesters per the needs or interests of your student; materials for each semester are independent of one another to allow flexibility.

Semester 1: Mathematics

Numbers surround us. Just try to make it through a day without using any. It's impossible: telephone numbers, calendars, volume settings, shoe sizes, speed limits, weights, street numbers, microwave timers, TV channels, and the list goes on and on.

The many advancements and branches of mathematics were developed through the centuries as people encountered problems and relied upon math to solve them. It's amazing how ten simple digits can be used in an endless number of ways to benefit man.

The development of these ten digits and their many uses is the fascinating story in *Exploring the World of Mathematics.*

Semester 2: Physics

Physics is a branch of science that many people condsider to be too complicated to understand. John Hudson Tiner puts this myth to rest as he explains the fascinating world of physics in a way that students can comprehend.

Did you know that a feather and a lump of lead will fall at the same rate in a vacuum? Learn about the history of physics from Aristotle to Galileo to Isaac Newton to the latest advances. Discover how the laws of motion and gravity affect everything from the normal activities of everyday life to launching rockets into space. Learn about the effects of inertia firsthand during fun and informative experiments.

Exploring the World of Physics is a great tool for students who want to have a deeper understanding of the important and interesting ways that physics affects our lives.

First Semester Suggested Daily Schedule

Date	Day	Assignment	Due Date	✓	Grade
		First Semester-First Quarter — Exploring the World of Mathematics			
Week 1	Day 1				
	Day 2	Read pages 4–12 • *Exploring the World of Mathematics* • (EWM)			
	Day 3				
	Day 4	Counting the Years - Questions **Mathematics Ch 1: Worksheet 1** • Page 21 • Lesson Plan • (LP)			
	Day 5				
Week 2	Day 6				
	Day 7	Read Pages 14–22 • (EWM)			
	Day 8				
	Day 9	Counting the Hours - Questions **Mathematics Ch 2: Worksheet 1** • Page 23 • (LP)			
	Day 10				
Week 3	Day 11				
	Day 12	Read Pages 24–28 • (EWM)			
	Day 13				
	Day 14	Read Pages 29–34 • (EWM)			
	Day 15				
Week 4	Day 16				
	Day 17	Muddled Measuring - Questions **Mathematics Ch 3: Worksheet 1** • Page 25 • (LP)			
	Day 18				
	Day 19	Read Pages 36–44 • (EWM)			
	Day 20				
Week 5	Day 21				
	Day 22	Measuring by Metric - Questions **Mathematics Ch 4: Worksheet 1** • Page 27 • (LP)			
	Day 23				
	Day 24	**Mathematics Chs 1–4: Quiz 1** • Page 93 • (LP)			
	Day 25				
Week 6	Day 26				
	Day 27	Read Pages 46–52 • (EWM)			
	Day 28				
	Day 29	Practical Mathematics - Questions **Mathematics Ch 5: Worksheet 1** • Page 29 • (LP)			
	Day 30				
Week 7	Day 31				
	Day 32	Practical Mathematics - Questions **Mathematics Ch 5: Worksheet 2** • Page 31 • (LP)			
	Day 33				
	Day 34	Read Pages 54–62 • (EWM)			
	Day 35				

Date	Day	Assignment	Due Date	✓	Grade
Week 8	Day 36				
	Day 37	The Greek Way with Math - Questions **Mathematics Ch 6: Worksheet 1** • Page 33 • (LP)			
	Day 38				
	Day 39	Read Pages 64–72 • (EWM)			
	Day 40				
Week 9	Day 41				
	Day 42	Names for Numbers - Questions **Mathematics Ch 7: Worksheet 1** • Page 35 • (LP)			
	Day 43				
	Day 44	Read Pages 74–82 • (EWM)			
	Day 45				
First Semester-Second Quarter — Exploring the World of Mathematics					
Week 1	Day 46				
	Day 47	Number Pattern - Questions **Mathematics Ch 8: Worksheet 1** • Page 37 • (LP)			
	Day 48				
	Day 49	**Chapters 5–8: Quiz 2** • Page 95 • (LP)			
	Day 50				
Week 2	Day 51				
	Day 52	Read Pages 84–94 • (EWM)			
	Day 53				
	Day 54	Endless Numbers - Questions **Mathematics Ch 9: Worksheet 1** • Page 39 • (LP)			
	Day 55				
Week 3	Day 56				
	Day 57	Endless Numbers - Questions **Mathematics Ch 9: Worksheet 2** • Page 41 • (LP)			
	Day 58				
	Day 59	Read Pages 96–106 • (EWM)			
	Day 60				
Week 4	Day 61				
	Day 62	Math for Scientists - Questions **Mathematics Ch 10: Worksheet 1** • Page 43 • (LP)			
	Day 63				
	Day 64	Read Pages 108–118 • (EWM)			
	Day 65				

Date	Day	Assignment	Due Date	✓	Grade
Week 5	Day 66				
	Day 67	Pure and Applied Math - Questions **Mathematics Ch 11: Worksheet 1** • Page 45 • (LP)			
	Day 68				
	Day 69	**Mathematics Chs 9–11: Quiz 3** • Page 97 • (LP)			
	Day 70				
Week 6	Day 71				
	Day 72	Read Pages 120–130 • (EWM)			
	Day 73				
	Day 74	Computing Machines - Questions **Mathematics Ch 12: Worksheet 1** • Page 47 • (LP)			
	Day 75				
Week 7	Day 76				
	Day 77	Computing Machines - Questions **Mathematics Ch 12: Worksheet 1** • Page 49 • (LP)			
	Day 78				
	Day 79	Read Pages 132–140 • (EWM)			
	Day 80				
Week 8	Day 81				
	Day 82	Bits and Bytes - Questions **Mathematics Ch 13: Worksheet 1** • Pages 51-52 • (LP)			
	Day 83				
	Day 84	Read Pages 142–152 • (EWM)			
	Day 85				
Week 9	Day 86				
	Day 87	Math on Vacation - Questions **Mathematics Ch 14: Worksheet 1** • Pages 53-54 • (LP)			
	Day 88				
	Day 89	**Mathematics Chs 12–14: Quiz 4** • Page 101 • (LP)			
	Day 90				
		Chapters 1-14: Test • Page 103 • (LP)			
		Mid-Term Grade			

Second Semester Suggested Daily Schedule

Date	Day	Assignment	Due Date	✓	Grade
		Second Semester-Third Quarter — Exploring the World of Physics			
Week 1	Day 91				
	Day 92	Read Pages 4–12 • *Exploring the World of Physics* • (EWP)			
	Day 93				
	Day 94	Motion - Questions **Physics Ch 1: Worksheet 1** • Page 57 • (LP)			
	Day 95				
Week 2	Day 96				
	Day 97	Read Pages 14–22 • (EWP)			
	Day 98				
	Day 99	Laws of Motion - Questions **Physics Ch 2: Worksheet 1** • Page 59 • (LP)			
	Day 100				
Week 3	Day 101				
	Day 102	Read Pages 24–32 • (EWP)			
	Day 103				
	Day 104	Gravity - Questions **Physics Ch 3: Worksheet 1** • Page 61 • (LP)			
	Day 105				
Week 4	Day 106				
	Day 107	Read Pages 34–40 • (EWP)			
	Day 108				
	Day 109	Simple Machines - Questions **Physics Ch 4: Worksheet 1** • Page 63 • (LP)			
	Day 110				
Week 5	Day 111				
	Day 112	**Physics Chs 1–4: Quiz 1** • Page 107 • (LP)			
	Day 113				
	Day 114	Read Pages 42–52 • (EWP)			
	Day 115				
Week 6	Day 116				
	Day 117	Energy - Questions **Physics Ch 5: Worksheet 1** • Page 65 • (LP)			
	Day 118				
	Day 119	Read Pages 54–64 • (EWP)			
	Day 120				

Date	Day	Assignment	Due Date	✓	Grade
Week 7	Day 121				
	Day 122	Heat - Questions **Physics Ch 6: Worksheet 1** • Page 67 • (LP)			
	Day 123				
	Day 124	Heat - Questions **Physics Ch 6: Worksheet 2** • Page 69 • (LP)			
	Day 125				
Week 8	Day 126				
	Day 127	Read Pages 66–76 • (EWP)			
	Day 128				
	Day 129	State of Matter - Questions **Physics Ch 7: Worksheet 1** • Page 71 • (LP)			
	Day 130				
Week 9	Day 131				
	Day 132	**Physics Chs 5–7: Quiz 2** • Page 109 • (LP)			
	Day 133				
	Day 134	Read Pages 78–88 • (EWP)			
	Day 135				
		Second Semester-Fourth Quarter — Exploring the World of Physics			
Week 1	Day 136				
	Day 137	Wave Motion - Questions **Physics Ch 8: Worksheet 1** • Page 73 • (LP)			
	Day 138				
	Day 139	Wave Motion - Questions **Physics Ch 8: Worksheet 2** • Page 75 • (LP)			
	Day 140				
Week 2	Day 141				
	Day 142	Read Pages 90–100 • (EWP)			
	Day 143				
	Day 144	Light - Questions **Physics Ch 9: Worksheet 1** • Page 77 • (LP)			
	Day 145				
Week 3	Day 146				
	Day 147	Light - Questions **Physics Ch 9: Worksheet 2** • Page 79 • (LP)			
	Day 148				
	Day 149	Read Pages 102–110 • (EWP)			
	Day 150				
Week 4	Day 151				
	Day 152	Electricity - Questions **Physics Ch 10: Worksheet 1** • Page 81 • (LP)			
	Day 153				
	Day 154	Read Pages 112–122 • (EWP)			
	Day 155				

Date	Day	Assignment	Due Date	✓	Grade
Week 5	Day 156				
	Day 157	Magnetism - Questions **Physics Ch 11: Worksheet 1** • Page 83 • (LP)			
	Day 158				
	Day 159	**Physics Chs 8–10: Quiz 3** • Page 111 • (LP)			
	Day 160				
Week 6	Day 161				
	Day 162	Read Pages 124–134 • (EWP)			
	Day 163				
	Day 164	Electromagnetism - Questions **Physics Ch 12: Worksheet 1** • Page 85 • (LP)			
	Day 165				
Week 7	Day 166				
	Day 167	Read Pages 136–142 • (EWP)			
	Day 168				
	Day 169	Nuclear Energy - Questions **Physics Ch 13: Worksheet 1** • Page 87 • (LP)			
	Day 170				
Week 8	Day 171				
	Day 172	Read Pages 144–152 • (EWP)			
	Day 173				
	Day 174	Future Physics - Questions **Physics Ch 14: Worksheet 1** • Page 89 • (LP)			
	Day 175				
Week 9	Day 176				
	Day 177	**Physics Chs 11–14: Quiz 4** • Page 115 • (LP)			
	Day 178				
	Day 179	**Physics Chs 1-14: Test** (Optional) • Page 117 • (LP)			
	Day 180				
		Final Grade			

Suggested Optional Science Lab:

There are a variety of companies that offer science labs that complement our courses. These items are only suggestions, not requirements, and they are not included in the daily schedule. We have tried to find materials that are free of evolutionary teaching, but please review any materials you may purchase. The following items are available from www.HomeTrainingTools.com.

Concepts of Math & Physics
The World of Physics
KT-PHSKIT Physics Workshop Kit

Special Projects

The Exploring series offers a unique perspective filled with biographical, historical, and scientific perspectives. By highlighting the work and relevance of scientists and innovators, students are introduced to the people behind the knowledge and discoveries that continue to impact their world. This provides exceptional learning opportunities above and beyond the worksheets, quizzes, and tests. Below are three areas of possible activities or bonus point projects that can be undertaken to enhance study.

Biographical

- Select your favorite scientist mentioned in the book and do a research paper on this person's life and/or work. Be sure to include details that enhance the understanding of why they worked in the area of science that they chose, information on their worldview (Christian or secular), and why their work remains relevant.
- There have been some amazing discoveries by women — see if you can find three discoveries by researching at your local library or online at parent-approved sites.

Historical

- Do three short essays — no more than two typed pages each — on discoveries that laid the groundwork for future science fields or the advancement of knowledge.
- Discover where 25 important discoveries related to mathematics or science took place; mark the map for each place and label with the name of each discovery.
- The Bible contains some amazing mathematical and scientific information. Using the geneaological information in Genesis 5, see if you can calculate how many years took place between creation and the Flood of Noah.

Scientific

- Imagine an invention related to mathematics or physical science that could change the way you and others live. See if you can visualize your invention by drawing it out or providing details that would enable someone else to understand the relevance of your invention and how it works.

Applied Learning

These ideas provide a way for the student to acquire knowledge and then apply it — whether that is done in a technical sense or by being able to recognize the concepts at work in the course of their daily experiences. Consider doing one of the two following options as an opportunity to earn bonus points or to extend the learning process:

- Take a spiral notebook and name it "My Learning Observations." Then, using the following concepts, mark the date and time you observe each example over a two-week period. Remember, science is happening around you all the time in every day life, so make sure your observations correlate with mathematics or physical science.
- You can keep a running study journal using the words and people to know on the following pages. By writing down the definition of words, or the contribution of an individual, you can develop a deeper understanding of the subject matter and have notes available when studying for quizzes and exams.

Exploring the World of Mathematics

People and Words to Know

Chapter 1 (p. 4–13)
Augustus Caesar
Bible
calendar
Gregorian calendar
Julian calendar
Julius Caesar

Chapter 2 (p. 24–23)
analog
census
Galileo
pi

Chapter 3 (p. 24–35)
BASIC computer language
metric system

Chapter 4 (p. 36–45)
Anders Celsius
celsius (thermometer)
decimals
fahrenheit
Kelvin scale
Isaac Newton

Chapter 5 (p. 46–53)
Ahmes
Great Pyramid of Giza
hypotenuse

Chapter 6 (p. 54–63)
ark of the covenant
conic sections
Elements of Geometry
ellipse
Euclid
golden ratio
hyperbola
Johannes Kepler
parabola
Pythagoras
Pythagorean theorem

Chapter 7 (p. 64–73)
abacus
Arabic numerals
Archimedes
Book of Calculating
byte
calculator
Fibonacci
Roman numerals

Chapter 8 (p. 74–83)
algebra
encrypted data
Eratosthenes
Fibonacci numbers
Carl Frederick Gauss
Internet
number theory
palindrome
prime number
sieve of Erathosthenes

Chapter 9 (p. 84–95)
bit
ENIAC
irrational number
rational number
square root

Chapter 10 (p. 96–107)
analytical geometry
Carl Louis Lindemann
Hooke's Law
Rene Descartes

Chapter 11 (p. 108–119)
applied mathematics
binomial
Leonhard Euler
factorial
Pierre de Fermat
Fermat's last theorem
four-color map problem
Königsberg bridge problem
Blaise Pascal
Pascal's triangle
peal (bell ringing)
permutations
probability
Andrew Wiles

Chapter 12 (p. 120–131)
analytical engine
Augusta Ada King (Countess of Lovelace)
Charles Babbage
Central Processing Unit (CPU)
difference engine
Gottfried Leibnitz
Howard H. Aiken
logarithm
John Napier
significant digits
slide rule
step reckoner

Chapter 13 (p. 132–141)
ASCII (American Standard Code for Information Interchange)
base two
binary number
compression routine
Grace Hopper
pixel

Chapter 14 (p. 142–153)
calculus
Cartesian coordinate system
Albrecht Dürer

Exploring the World of Physics

People and Words to Know

Chapter 1 (p. 4–13)
energy
heat
matter
nuclear energy
Galileo
Aristotle
current
intensity
pendulum
Leaning Tower of Pisa
resistance
Apollo 15
David Scott
Robert Boyle
vacuum
gravity
force
friction
velocity
acceleration
parabola
projectile motion

Chapter 2 (p. 14–23)
first law of motion
second law of motion
force equation
third law of motion
impulse
Robert Goddard
conservation of momentum
vector
Johannes Kepler

Chapter 3 (p. 24–33)
ellipse
first law of planetary motion
second law of planetary motion
third law of planetary motion
bubonic plague
center of gravity
density
center of mass
Principia
global positioning system

Chapter 4 (p. 34–41)
incline plane
pulley
simple machines
fulcrum
Sicily
Great Pyramid of Giza
windlass
block and tackle

Chapter 5 (p. 42–53)
James Prescott Joule
joule
foot-pounds
British thermal units
James Watt
Bryan Allen
Gossamer Albatross
law of conservation of energy

Chapter 6 (p. 54–65)
heat capacity
specific heat
Josiah Wedgwood
pyrometer
thermistor
Anders Celsius
Antoine Lavoisier
caloric
Count Rumford
John Dalton
atomic theory of matter
Robert Brown
steam engine
thermodynamics
Kelvin
Nicolas Carnot
entropy
first law of thermodynamics
heat pumps
law of entropy
second law of thermodynamics
third law of thermodynamics

Chapter 7 (p. 66–77)
standard pressure
standard temperature
elasticity
elastic limit
McPherson struts
Hooke's law
Robert Hooke
Christopher Wren
Edmund Halley
Great Fire of London
hydraulic press
Pascal's principle
Blaise Pascal
buoyancy
law of buoyancy
Jacques Charles
ideal gas law
Joseph Gay-Lussac
diffusion
Thomas Graham
kinetic theory of gases
Bernoulli's principle

Chapter 8 (p. 78–89)
transverse
displacement
wave equation
longitudinal
vocal cords
node
overtones
sympathetic vibration
reverberation
SONAR
sonogram
ultrasonic
voiceprint
Alexander Graham Bell
Fourier analysis
Joseph Fourier

sonic boom
Christian Doppler
Doppler effect

Chapter 9 (p. 90–101)
prism
convex lens
iris
pupil
real image
cornea
retina
colorblindness
cones
primary colors
cyan
mirage
optical illusion
concave mirror
law of reflection
reflecting telescope
virtual image
convex mirror
focal length
magnification
Anton Leeuwenhoek
Jean Foucault

Chapter 10 (p. 102–111)
static electricity
Thales of Melitus
William Gilbert
electron
amber
Charles Du Fay
Isaac Newton
proton
Benjamin Franklin
Charles Coulomb
inverse square law
law of gravity
mass
rods
insulators
nonconductors
conductor
circuit
aluminum
Andrè Ampére
Lunar Rover
Count Volta
hybrid cars
Georg Simon Ohm
voltage
Samuel F.B. Morse
pressure
heat energy
Ohm's law
transformer

Chapter 11 (p. 112–123)
magnetite
north pole
south pole
lodestone
compass
North Star
Polaris
Robert Norman
magnet
axis
geographic poles
magnetic north
magnetic variation
north geographic pole
northern Lights
south geographic pole
magnetic field
lines of force
alnico
cobalt
gold
magnesium
paramagnetic
three laws of magnetism
ferromagnetic
magnetic domains
nucleus
temporary magnet
Marie Curie
permanent magnet
Pierre Curie
cow magnets
Curie temperature
horseshoe magnet
Hans Christian Oersted
electromagnetism
Joseph Henry
William Sturgeon
electromagnet
electromagnetic induction
Michael Faraday
atomic fusion
circuit breaker
magnetic levitation
speed of light
Polarized light

Chapter 12 (p. 124–135)
Cavendish Laboratory
James Clerk Maxwell
infrared light
Maxwell's field equations
spectrum
ultraviolet
electromagnetic waves
frequency
amplitude
Guglielmo Marconi
Rudolf Hertz
Albert Einstein
microwaves
radiation
x rays
interference
wavelength
amplitude modulation
bandwidth
frequency modulation
ionosphere
gamma rays
Wilhelm Roentgen
Archimedes
Ernest Rutherford
first scientific revolution
second scientific revolution
special theory of relativity
photoelectric effect
Brownian motion
general theory of relativity
Arthur Holly Compton
photon
chain reaction
cosmic rays

Manhattan Project
nuclear chain reaction
Charles Wilson
cloud chamber
Compton effect
momentum

Chapter 13 (p. 136–143)

neutron
carbon-14
quarks
uranium-235
uranium-238
alpha particle
nuclear fusion
Enrico Fermi
Lise Meiter
yellowcake
plutonium
breeder reactor
moderator
fusion reaction
kinetic energy
mass energy equation
cold fusion

Chapter 14 (p. 144–153)

ultraviolet catastrophe
proportionality constant
quantum
Max Planck
Planck's constant
ground state
Niels Bohr
Prince Louis de Borglie
fundamental frequency
Thomas Young
Simon Laplace
Heisenberg uncertainty principle
Werner Heisenberg
quantum mechanics
resonance
absolute zero
cryogenic
nanotechnology
superconductors
Kitt Peak
acoustics
conduction
convection
Daniel Fahrenheit
decibels
lever
mechanical advantage
pitch
potential energy
watt
wheel and axle

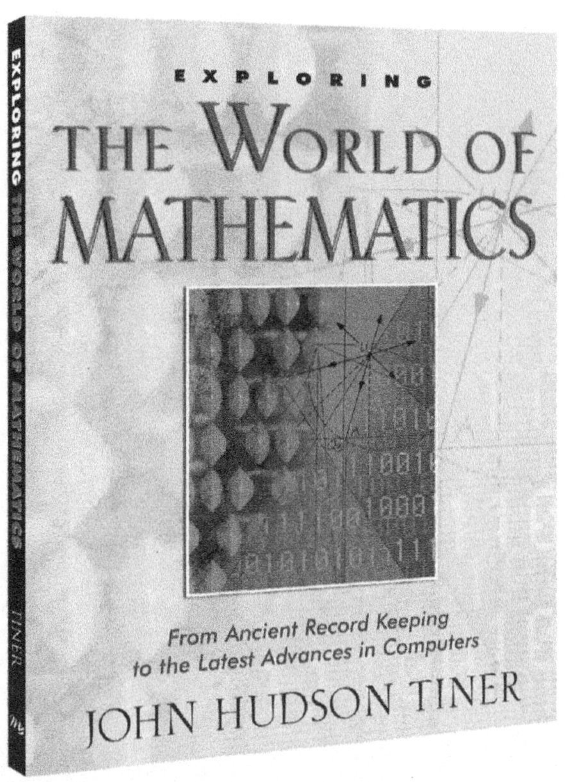

Mathematics Worksheets

for Use with

Exploring the World of Mathematics

Answer T or F for true or false, fill in the blank, or select the letter for the phrase that best completes the sentence.

T (F) 1. The extra day, or leap day, every four years was put in the calendar to honor Augustus Caesar.

T (F) 2. The Gregorian calendar has 100 leap days every 400 years.

3. What is the main reason to have leap days? To mach the seasons

A B C D 4. The first calendar with a leap day every four years was the one
(A.) authorized by Julius Caesar
B. used by the American colonies after 1752
C. used by the Babylonians
D. used by the Egyptians

Matching

5. C day	a. due to the tilt of the earth's axis, equal to three months
6. E week	b. earth revolves around the sun once
7. D month	c. earth rotates on its axis once
8. A season	d. moon revolves around the earth once
9. B year	e. seven days

Try Your Math

10. The Bible says that Methuselah died at age 969 years (Gen. 5:27). What would be that age in days? (Ignore leap years.) 353,685

11. Using the Babylonian calendar of 360 days in a year, how many days are in one-third of a year; one-fifth of a year; one-twentieth of a year; one-sixtieth of a year?
120, 72, 18, 6

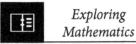

12. Find the population of your city and calculate how many people are likely to have a birthday on February 29. .7 %

Answer T or F for true or false, or select the letter for the phrase that best completes the sentence.

A B C D 1. The shortest naturally occurring period of time that ancient people could observe was the (A. day, B. hour, C. week, D. year).

T F 2. The Egyptians divided daylight into 8 or 12 hours depending on whether it was winter or summer.

A B C D 3. The inventors of the hourglass were the (A. Babylonians, B. British Navy, C. Egyptians, D. Romans).

A B 4. A watch with a sweep second hand is known as (A. an analog, B. a digital) watch.

T F 5. Meridians are imaginary lines going around the earth parallel to the equator.

A B C D 6. Military time has hours numbered from 0000 to (A. 0400, B. 1200, C. 2400, D. 3600).

A B C D 7. Time zones were introduced when it became common to travel by (A. airplanes, B. ox carts, C. ships, D. trains).

A B 8. The International date line is in the (A. Atlantic, B. Pacific) ocean.

A B 9. Atomic clocks proved that the earth's rotation (A. is, B. is not) uniform.

A B C 10. The United States became an independent nation in 1776. In 1976, the country celebrated the fact that the United States was two (A. decades, B. centuries, C. millenniums) old.

Try Your Math

11. Assume that the first four-hour watch began at midnight. What time would it be at five bells on the second watch? 6:30 am

12. Feel your pulse at the wrist and count the number of beats in a minute. Calculate the number of times your heart beats in a day.

13. An office job is often described as working from 9 to 5. This means 9:00 a.m. to 5:00 p.m. How many hours is this? 8

14. At 4:00 p.m., a family on vacation drives from Mountain Standard Time into Central Standard Time. Should their watches be set one hour earlier to 3:00 p.m. or one hour later to 5:00 p.m.? one hour later

Answer T or F for true or false, fill in the blank, or select the letter for the phrase that best completes the sentence.

A B C D 1. NASA's Climate Orbiter to Mars failed because (A. American and French engineers did not communicate with one another, B. engineers used two different measures of force, C. fuel had been measured improperly, D. the spacecraft weighed too much).

A B 2. A troy ounce was used to measure (A. small and expensive, B. large and inexpensive) items.

A B C D 3. A scruple was a standard of weight for measuring (A. barley, B. diamonds, C. drugs, D. potatoes).

A B C D 4. At first, the United States Customary system agreed with that of (A. Britain, B. France, C. Morocco, D. Spain).

T F 5. The American ton and the British tonne are identical in weight.

A B C D 6. Most early measures of distance were based on (A. animal strides, B. human body, C. parts of ships, D. Roman military terms).

7. The length of a mile in feet is 5,280.

8. "A pint is a Pound the world around."

Choose the larger:

A B 9. A. foot, B. yard

A B 10. A. fathom, B. yard

A B 11. A. nautical mile, B. statute mile

A B 12. A. cup, B. quart

A B 13. A. bushel, B. peck

Try Your Math

14. Recall that a hand is four inches. How tall is a horse in inches that is 15 hands tall? How tall in feet? 60 inches

15. Change your weight from pounds to ounces.

16. The tallest mountain on earth is Mt. Everest. Its summit is 29,035 feet above sea level. How high is the mountain in miles? 5.5 miles

Answer T or F for true or false, or
select the letter for the phrase that best completes the sentence.

A B 1. The metric system began in (A. Britain, B. France).

A B C 2. The metric system is based on powers of (A. 2, B. 10, C. 12).

T F 3. The metric system was designed specifically to meet the needs of merchants.

A B C D 4. Currently, the meter is defined as (A. 1,640,763.73 wavelengths of krypton gas, B. 1/10,000,000 of the distance from the equator to the North Pole, C. the distance between two scratch marks on a metal rod, D. the distance light travels in 1/299,792,458 of a second).

T F 5. Volume (capacity) is a derived unit because it is based on a container that is 1/10 of a meter on each side.

A B C 6. One meter is slightly longer than one (A. inch, B. yard, C. mile).

A B C 7. One liter is slightly larger than one (A. pint, B. quart, C. gallon).

A B C 8. One kilogram is about 2.2 times as much as (A. one ounce, B. one pound, C. one ton).

A B 9. A standard kilogram is defined by (A. the mass of a platinum cylinder, B. the wavelength of krypton gas).

A B C D 10. Daniel Fahrenheit set the boiling temperature of water on his thermometer at (A. 0, B. 32, C. 100, D. 212) degrees.

A B C D 11. Most people liked Fahrenheit thermometers because (A. they were free, B. they were accurate, C. Fahrenheit was an Englishman, D. daytime temperatures stayed between 0 and 100 degrees).

T F 12. The metric system is illegal to use in the United States.

A B C 13. The (A. Celsius, B. Fahrenheit, C. Kelvin) temperature scale starts at absolute zero.

Answer T or F for true or false, or
select the letter for the phrase that best completes the sentence.

A B C D 1. The country known as "the gift of the Nile" was (A. Burundi, B. Egypt, C. Rwanda, D. Sudan).

T F 2. One of the reasons the Egyptians developed mathematics was to figure taxes.

T F 3. Geometry means "to measure a pyramid."

A B 4. The long side of a right triangle is known as the (A. leg, B. hypotenuse).

A B C D 5. The Egyptian knotted rope was used to measure out (A. a pyramid with sloping sides, B. a rectangle with parallel sides, C. a silo of a fixed height, D. a triangle with a right angle).

T F 6. A quadrilateral and triangle are two names for the same figure.

A B C 7. Square feet is an example of a measure of (A. area, B. distance, C. volume).

A B C D 8. Doubling the length, width, and height of a box gives it a volume (A. twice as great, B. three times as great, C. six times as great, D. eight times as great).

A B C D 9. The distance around a circle is called its (A. circumference, B. diameter, C. height, D. radius).

A B 10. The expressions 22 to 7, 22/7, 3.14, and π all refer to the ratio of the circumference of a circle to its (A. diameter, B. radius).

Matching

11. E circle
12. A pentagon
13. D rectangle
14. C right triangle
15. B square

a. a polygon with five sides
b. a rectangle with four equal sides
c. a polygon with three sides and one right angle
d. a quadrilateral with opposite sides parallel and equal in length
e. is not a polygon

Matching Formulas

16. D 2L + 2W
17. E 4S
18. F Ah
19. B L x W
20. A πr^2
21. C S^2

a. area of a circle
b. area of a rectangle
c. area of a square
d. perimeter of a rectangle
e. perimeter of a square
f. volume

22. A room is 10 feet wide and 14 feet long. How many square tiles, one foot on a side, would be needed to completely cover the room? 140 tiles

Answer T or F for true or false, fill in the blank or select the letter for the phrase that best completes the sentence.

A B 1. The (A. Egyptians, B. Greeks) strove to understand the principles of mathematics.

2. The sum of the Squares of the legs of a right triangle are equal to the Square of the hypotenuse.

A B C 3. The figure that encloses the greatest area with the least perimeter or circumference is the (A. circle, B. square, C. triangle).

A B C D 4. A whispering gallery has a shape like (A. a circle, B. a hyperbola, C. a parabola, D. an ellipse).

A B 5. If an object follows an elliptical orbit, then it is on (A. a closed, B. an open) path.

Matching

6. E Archimedes
7. D Euclid
8. C Johannes Kepler
9. A Pythagoras
10. B Thales

a. discovered that the sum of the three angles of any triangle is 180 degrees

b. used ratios to find the heights of buildings

c. proved planets follow elliptical orbits

d. wrote *Elements of Geometry*

e. ancient Greek who worked out a way to show large numbers that he called myriads

Matching

11. B circle — a. a mirror of this shape will focus sunlight
12. D ellipse — b. all points are the same distance from the center
13. A parabola — c. the first part of the name means over or beyond
14. C hyperbola — d. the orbit of Halley's comet is of this shape

Answer T or F for true or false.

T F 1. Some cultures counted with 20 as the base.

T F 2. Any mark that is used to stand for a number is called a digit.

T F 3. The value of a number does not depend on how it is represented.

T F 4. Place value gives a symbol a different value depending upon its location.

T F 5. The digit 0 was invented at the same time as the digits 1 through 9.

T F 6. The digit 0 first came into use in India.

T F 7. Italian merchants packaged goods by the dozen because the number 12 could be divided into smaller portions.

T F 8. Mathematics is sometimes called the ruler of science.

T F 9. Isaac Newton introduced the use of place value and the numeral 0 to Europe.

T F 10. Fibonacci wrote a book called *Elements of Counting.*

T F 11. The prefix bi means one-half.

T F 12. The word billion has the same meaning in England as in the United States.

T F 13. Of the prefixes giga, mega, and tera, the one that has the greatest value is mega.

Answer T or F for true or false, fill in the blank, or select the letter for the phrase that best completes the sentence.

A B C D 1. The sentence, "Madam, I'm Adam" is an example of a (A. composite statement, B. palindrome, C. permutation, D. Roman oration).

A B C D 2. The study of the properties of whole numbers is called (A. algebra, B. geometry, C. number theory, D. real analysis).

A B 3. The general form of an even number is (A. 2n, B. 2n + 1), with *n* a whole number.

A B 4. Two is a factor of all (A. even, B. odd) numbers.

A B 5. The number with the greater number of factors is (A. 12, B. 13).

A B C 6. Nine is an example of (A. a prime, B. an even, C. an odd) number.

A B 7. An example of a composite number is (A. 11, B. 12).

A B C D 8. Prime numbers can be found with the sieve of (A. Eratosthenes, B. Euclid, C. Gauss, D. Pythagoras).

A B 9. As you count higher and higher, prime numbers become (A. more and more common, B. rarer and rarer).

T F 10. The statement "A composite number can be written as the product of prime numbers in only one way" has not yet been proven to be true.

T F 11. The statement "Every even number greater than two is the sum of two primes" has not yet been proven to be true.

A B 12. Encrypted data (A. is especially easy for any computer to read and display, B. can only be read by the sender and receiver).

A B C D 13. Fibonacci numbers could also be called (A. calculating with an abacus, B. the problem of adding Adder snakes, C. the problem of multiplying rabbits, D. the problem of the Leaning Tower of Pisa).

14. The next Fibonacci number after 89 and 144 is 233.

T F 15. The Fibonacci series of numbers is seldom found in nature.

Matching

16. B 1881, 121, 1001 — a. Fibonacci numbers
17. C 2, 3, 5, 7, 11, 13 . . . — b. palindromes
18. A 1, 1, 2, 3, 5, 8, 13 . . . — c. prime numbers
19. D 1, 4, 9, 16, 25 . . . — d. square numbers
20. E 1, 3, 6, 10, 15 . . . — e. triangular numbers

Answer T or F for true or false, or select the letter for the phrase that best completes the sentence.

A B 1. Another name for whole numbers is (A. irrational numbers, B. integers).

A B 2. Mathematicians rely on (A. examples, B. proofs) to show whether a statement is true or false.

T F 3. The sum of two whole numbers is always a whole number.

T F 4. The product of two whole numbers is always a whole number.

T F 5. The quotient of two whole numbers is always a whole number.

T F 6. A plus sign, +, can mean both the operation of addition and that a number is a positive integer.

A B C D 7. Before the invention of calculators, shares were used to reduce the necessity of doing (A. addition, B. subtraction, C. multiplication, D. division).

A B C D 8. The American colonies divided the real, a Spanish coin, into (A. 2, B. 4, C. 8, D. 12) pieces.

A B C D 9. Two bits is equal to (A. 12½, B. 25, C. 50, D. 100) cents.

A B C D 10. A common fraction can be changed into a decimal by dividing the numerator by the (A. denominator, B. greatest common factor, C. least common multiple, D. remainder).

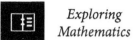

Answer T or F for true or false, or select the letter for the phrase that best completes the sentence.

A B 11. The expression ⅔ = 0.666 . . . is an example of a (A. repeating, B. terminating) decimal.

T F 12. Every number can be written as the ratio of two whole numbers.

A B C D 13. The square root of two, is an example of (A. a common fraction, B. an irrational number, C. a rational number, D. a terminating decimal).

A B C D 14. The digits of the square root of two, $\sqrt{2}$, when expressed as a decimal (A. do not repeat, B. do not terminate, C. do not form a pattern, D. all of the above).

Matching

15.__A__ 1, 2, 3, 4, 5 — a. counting numbers

16.__G__ 0, 1, 2, 3, 4, 5 — b. even numbers

17.__B__ 2, 4, 6, 8, 10 — c. integers

18.__E__ 1, 3, 5, 7, 9 — d. irrational numbers

19.__C__ -3, -2, -1, 0, +1, +2, +3 — e. odd numbers

20.__F__ $1/1$, ½, ¾, ⅔ — f. rational numbers

21.__D__ $\sqrt{2}$, π, $(1 + \sqrt{5})/2$ — g. whole numbers

Answer T or F for true or false, or select the letter for the phrase that best completes the sentence.

T F 1. Normally, a variable can have more than one value.

A B C D 2. Discovering the value of x when y is equal to zero is called (A. modernizing, B. normalizing, C. solving, D. zeroing) the equation.

A B C D 3. The 5 in the equation $5x + 2 = 0$ is called (A. a coefficient, B. a constant, C. an equalizer, D. a variable).

A B 4. An equation such as $ax^2 + bx + c = 0$ is called an equation of the (A. first, B. second) degree.

T F 5. An equation of the type $y = kx$ is called a linear equation.

T F 6. A scientific law must be stated in the metric system of units to be true.

A B C D 7. The expression $1/x$ is called the (A. base, B. identify, C. reciprocal, D. square root) of x.

A B C D 8. In the 1600s, the French mathematician René Descartes discovered a way to combine algebra with (A. computer programming, B. geometry, C. number theory, D. physics).

A B C 9. The three problems that had resisted solutions since ancient Greek times were trisecting the angle, doubling a cube, and (A. bisecting an angle, B. making a right angle, C. squaring the circle).

Match the equation with the figure on right.

10._____ $y = kx$

11._____ $y = kx^2$

12._____ $y = k\sqrt{x}$

13._____ $y = k/x$

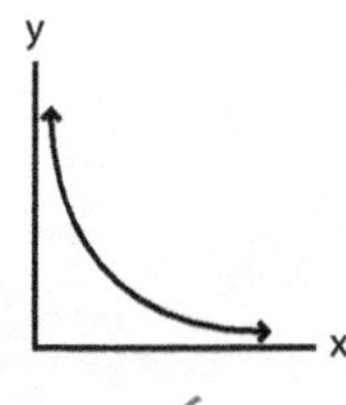

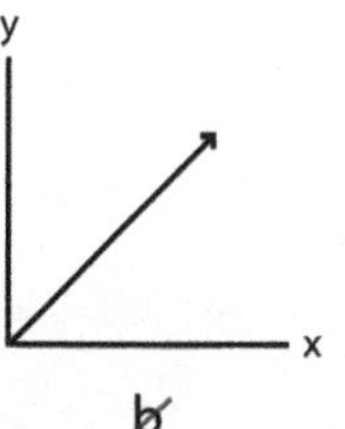

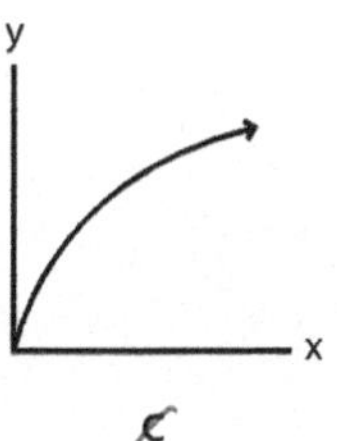

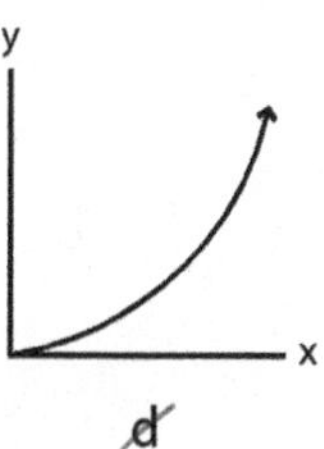

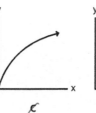

Match the statement with the figure above.

14._____ The length a spring stretches (y-axis) is directly proportional to the force pulling on the spring (x-axis).

15._____ The distance an object falls (y-axis) under the influence of gravity is directly proportional to the square of the time it has fallen (x-axis).

16._____ The volume of a gas (y-axis) is inversely proportional to the pressure acting on the gas (x-axis).

17._____ The period of a pendulum (y-axis) is directly proportional to the square root of the length of the pendulum (x-axis).

Answer T or F for true or false, or select the letter for the phrase that best completes the sentence.

A B 1. Mathematics for practical use is called (A. applied, B. pure) mathematics.

A B 2. Discovering large prime numbers to encode data is an example of (A. applied, B. pure) math.

A B C D 3. The problem that a computer helped solve was the (A. bell peal problem, B. binomial theorem, C. four-color map problem, D. Königsberg bridge problem).

A B 4. Each arrangement of the letters ABC, ACB, BAC, BCA, CAB, and CBA is called a (A. combination, B. permutation).

A B C D 5. The expression 3! is read as "three (A. combinations," B. factorial," C. permutations," D. probabilities").

A B C D 6. The value of 5! is (A. 24, B. 25, C. 120, D. 125).

A B 7. To calculate the number of ways that items can be arranged, (A. add, B. multiply) the number of choices for each position.

T F 8. The study of combinations and permutations has no application in everyday life.

Matching

9. E Andrew Wiles
10. B Blaise Pascal
11. A Isaac Newton
12. D Leonhard Euler
13. C Pierre de Fermat

a. Discovered how to calculate the coefficients of a binomial raised to a power.

b. He called his triangle an arithmetic triangle.

c. His last theorem was solved in 1995.

d. Solved the Königsberg bridge problem.

e. Proved that $x^n + y^n = z^n$ has no solution with whole numbers except for n = 2.

Try Your Math

14. The state of Missouri has license plates with three letters followed by three digits. How many license plates are possible? 17,576,000

Answer T or F for true or false, fill in the blank, or select the letter for the phrase that best completes the sentence.

A B C D 1. Any number raised to the zero power is (A. 0, B. 1, C. 2, D. undefined).

T F 2. Babbage completed the analytical engine shortly before his death.

T F 3. Powers of ten can be multiplied by adding their exponents.

T F 4. Fractional exponents are not allowed.

T F 5. A logarithm is an exponent.

6. The expression $\log_{10}3 = 0.477$ is read as "The logarithm of the number __3__ in base __10__ is __0.477__."

A B C D 7. The number 5,280 changed to standard notation is (A. .5280 x 10^1, B. 5,280 x 10^3, C. 5.28 x 10^2, D. 5.28 x 10^3).

A B C 8. Scientific measurements are (A. less accurate, B. no more accurate, C. significantly more accurate) than the instruments used to make the measurements.

T F 9. A slide rule multiplies two numbers by adding their logarithms.

A B C D 10. In the early days of computers, input was mainly by (A. colored ribbons, B. punched cards, C. spoken commands, D. switches and relays).

11. In this list, which one is considered the "heart" of a computer: input, control program, memory, central processing unit, output. central processing unit

A B C D 12. The letters RAM stand for (A. random access memory, B. reasonably accurate member, C. recent abacus modification. D. Robert A. Morley).

Matching

13.__H__ Augusta Ada Byron, Lady Lovelace

14.__C__ Howard H. Aiken

15.__D__ Charles Babbage

16.__F__ Herman Hollerith

17.__G__ Johannes Kepler

18.__A__ Gottfried Leibnitz

19.__E__ John Napier

20.__B__ Blaise Pascal

a. built a calculator called the Step Reckoner

b. built a calculator to help his father, a tax collector

c. built the first general-purpose calculating machine

d. built the difference engine

e. invented logarithms

f. invented tabulating machines used in the 1890 census

g. spent six years calculating the orbit of Mars

h. wrote the first computer program

Answer T or F for true or false, fill in the blank, or select the letter for the phrase that best completes the sentence.

1. Base 10 uses the digits 0 through 9.

T F 2. Base 2 uses the digits 0, 1, and 2.

A B C D 3. In computer usage, a single position for a binary digit is called a (A. bit, B. byte, C. kilo, D. pixel).

A B C 4. A single bit can have (A. one, B. two, C. ten) different value(s).

A B C D 5. In personal computers, a byte of data is made of (A. one, B. two, C. eight, D. ten) bit(s).

T F 6. Each character written in ASCII code takes a byte to represent it.

A B C 7. A pixel in a (A. black and white photograph, B. color photograph, C. line drawing) requires the greatest number of bytes.

A B C D 8. In 1861, James Clerk Maxwell made a color photograph using (A. computer enhancement, B. color ink drops sprayed on paper, C. polarized light, D. the three colors of red, green, and blue).

T F 9. Text files cannot be compressed.

T F 10. A pixel is always equal to one bit.

A B C D 11. Video images can be compressed by (A. converting black and white images to color images, B. having reporters avoid standing in front of a blue sky, C. transmitting all pixels that are the same as the previous one, D. transmitting only those pixels that are different from the previous one).

A B C D 12. Moore's law states that computers double in power every 18 (A. days, B. months, C. decades, D. years).

A B C D 13. The bug that Grace Hopper found in the Mark II computer turned out to be (A. a hardware problem, B. a moth caught between mechanical relays, C. a software problem, D. a problem caused by human error).

A B C D 14. A computer with components put farther apart will run more slowly because (A. electric signals can go no faster than the speed of light, B. larger components must be made of less costly materials, C. of resistance in the wires, D. the electrons get lost).

A B C D 15. Engineering students at MIT in the 1950s answered simple questions in computer science with (A. mechanical calculators, B. model cars, C. model trains, D. radio-controlled airplane).

T F 16. The binary digit 1 stands for true, yes, on, or is possible.

A B 17. The query that is most likely to result in more citations is (A. an OR query, B. an AND query).

Try Your Math

18. The Constitution of the United States has 4,609 words and 26,747 characters. At the rate of 7,000 bytes per second, how long would it take a computer to download the Constitution of the United States as an uncompressed text file? 3.8 seconds

PUZZLES

Puzzle 1: What is so special about 142,857?

The number 142,857 gives interesting results when multiplied by 1 through 6:

1 x 142,857 = 142,857
2 x 142,857 = 285,714
3 x 142,857 = 428,571
4 x 142,857 = 571,428
5 x 142,857 = 714,285
6 x 142,857 = 857,142

The digits in the answer cycle through the digits 142,857 in the same order. You might predict that the answer to 7 x 142,857 would have the same digits. Such a prediction would be wrong. Multiply 7 x 142,857 and see the surprising answer. 999,999

Puzzle 2: Multiplying by 99

Some numbers are fun to play around with.

2 x 99 = 198
3 x 99 = 297
4 x 99 = 396
5 x 99 = 495
6 x 99 = 594
7 x 99 = 693
8 x 99 = 792
9 x 99 = 891

The left-most digit (the one in the 100s place) in the answer goes from 1 to 8 while the right-most digit (the one in the 1s place) goes from 8 to 1. Try to figure out the reason for the pattern.

Puzzle 3: On the Road to St. Ives

Try to solve this people on the road puzzle that was turned into an English children's rhyme:

As I was going to St. Ives
I met a man with seven wives;
Every wife had seven sacks;
Every sack had seven cats;
Every cat had seven kits [kittens];
Kits, cats, sacks, and wives,
How many were going to St. Ives?

(Can you figure out the answer to the riddle?)
one

Puzzle 4: Send More Money

Here is an addition problem with letters taking the place of numbers. Solve the problem by replacing each letter with one of the digits 0 through 9. Use the same digit for the same letter throughout. In a puzzle like this, it is understood that 0 is not allowed as the first letter in any of the words.

```
   S E N D
 + M O R E
 M O N E Y
```

Hint: Start on the left side. M in MONEY must be 1 because even with a carry, the sum of S and M is less than 20. 10,652

Puzzle 5: The 3N + 1 Problem

Pick a number, divide by 2 if it is even. But if it is odd, then multiply by 3 and add 1. Keep on doing this to see where it leads.

For instance, start with 6 (even)

Divide: 6 ÷ 2 = 3 (odd)
Multiply by 3 and add 1: 3 x 3 + 1 = 10 (even)
Divide: 10 ÷ 2 = 5 (odd)
Multiply by 3 and add 1: 3 x 5 + 1 = 16 (even)
Divide: 16 ÷ 2 = 8 (even)
Divide: 8 ÷ 2 = 4 (even)
Divide: 4 ÷ 2 = 2 (even)
Divide: 2 ÷ 2 = 1 (stop)

In every number that mathematicians have tried, the result is always 1. However, no one has yet been able to supply a proof or find an example that does not end with one. Try it with 18.

Puzzle 6: Samson's Riddle

The Bible has puzzles such as Samson's riddle in Judges 14:14: He replied, "Out of the eater, something to eat; out of the strong, something sweet." Hint: You can find the answer in Judges 14:8.

Puzzle 7: Sock Puzzle

Because of an electrical power failure, a boy must get dressed in a dark bedroom. His sock drawer has 10 blue socks and 10 black socks, but in the darkness he cannot tell them apart. He dresses anyway. He reaches into the drawer to grab spare socks so he can change into matching colors later. How many should he take to be certain he has a matching pair?

Puzzle 8: River Crossing

A canoeist must cross a river with three things, but his canoe can hold only one thing at a time. How can the canoeist get a wolf, goat, and carrots across a river? If left alone, the wolf would eat the goat, and the goat would eat the carrots.

Puzzle 9: Durer's Number Square

You can try your hand at making a number square by using the digits 1 through 9 in a three by three square. Each of the rows, columns, and diagonals should add to the same number. Eight different squares are possible.

Puzzle 10: Grass to Milk

Here is a problem to work on a calculator. The answer, when held upside down, shows the name of a four-legged animal that can change green grass into white milk. Find the product of the prime numbers 7, 17, and 23. Add to that answer the area in square feet of a field that is 201 feet on a side, the number of seconds in a day, and 186,000 miles per second (the speed of light). In adding these numbers, ignore the units. Now turn the calculator display around. What name do you see?

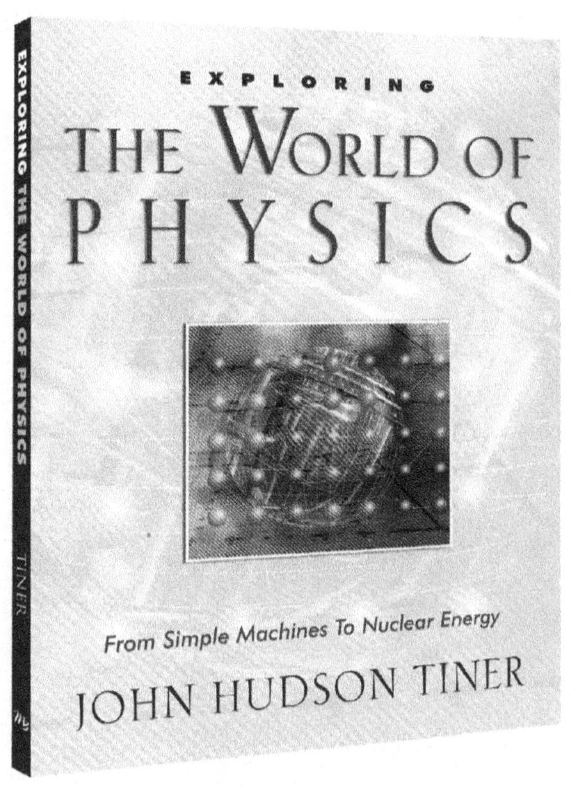

Physics Worksheets

for Use with

Exploring the World of Physics

Answer T or F for true or false, fill in the blank, or select the letter for the phrase that best completes the sentence.

A B C D 1. Physics is the science that explores how energy acts on (A. heat, B. light, C. matter, D. sound).

T F 2. The ancient Greeks were noted for their careful experiments.

T F 3. The regular back and forth motion of a pendulum was used to regulate the first accurate clocks.

T F 4. In Galileo's time, only length and time could be measured with any accuracy.

A B C D 5. A feather and lump of lead will fall at the same speed in (A. a high speed wind tunnel, B. the atmosphere, C. a vacuum, D. water).

6. To calculate speed, divide distance by time.

A B C D 7. To study the motion of falling objects, Galileo (A. beat them into cubes, B. dropped them from a high tower, C. pushed them from a cliff, D. rolled them down a ramp).

A B C D 8. Acceleration is found by dividing the (A. average velocity, B. distance, C. gravity, D. change in speed) by the change in time.

A B 9. On earth, the acceleration due to gravity is (A. 32 ft/sec^2, B. 60 miles/hour).

For More Study

10. Suppose a canoeist takes 70 days to paddle the entire length of the Mississippi River, a distance of 3,710 miles. The canoeist's average speed in miles per day is 53.

11. An ordinary passenger car can accelerate to 60 miles per hour in about eight seconds. What is the car's acceleration? 7.5 mi./hr

12. On the moon, the acceleration due to gravity is 5.3 ft/sec^2 rather than 32 ft/sec^2. If an object fell six seconds before hitting the ground, it strikes the ground with a speed of 31.8 ft/sec. (Hint: Use the final velocity equation.)

Answer T or F for true or false, fill in the blank, or select the letter for the phrase that best completes the sentence.

T F 1. Velocity and speed mean the same.

T F 2. A force must act on an object to put the object in motion, give it greater speed, slow it, or change its direction.

T F 3. All objects come to a stop unless some force keeps them going.

T F 4. A ball rolling on a flat surface comes to a stop because of the force of (A. friction, B. gravity).

A B C 5. Isaac Newton's first law of motion was based on experiments done by (A. Aristotle, B. Galileo, C. Newton, himself).

A B C D 6. Inertia is a property of matter that resists changing its (A. electric charge, B. mass, C. momentum, D. velocity).

T F 7. Only very massive objects have inertia.

T F 8. Acceleration is any change of speed or direction.

9. State the second law of motion.

10. State the third law of motion.

11. Momentum is the mass of an object times its velocity.

T F 12. The law of conservation of momentum is one of the most firmly established laws of science.

Matching

13. E	first law of motion	a. $a = f/m$
14. A	second law of motion	b. $f = m \times a$
15. C	third law of motion	c. $f_{ab} = -f_{ba}$
16. B	force equation	d. $I = f \times t$
17. D	definition of impulse	e. If $f = 0$ then $a = 0$
18. F	definition of momentum	f. $p = m \times v$

Answer T or F for true or false, fill in the blank, or select the letter for the phrase that best completes the sentence.

T F 1. During Kepler's time, most people believed the laws governing motions in the heavens differed from those for motions on earth.

A B C D 2. Kepler proved that planets traveled in orbits that were (A. circular, B. elliptical, C. parabolic, D. straight-line).

A B 3. A planet travels (A. faster, B. slower) when closer to the sun.

4. State the second law of planetary motion.

the straight line joining a planet with the sun sweeps out equal areas in equal intervals of time

T F 5. Kepler's third law of motion reveals that planets farther from the sun take longer to orbit the sun.

A B 6. Isaac Newton built upon the discoveries of Galileo and (A. Aristotle, B. Kepler).

T F 7. Isaac Newton came from a rich and powerful family.

T F 8. The direction the force of gravity acts on the moon is toward the center of the earth.

A B C D 9. The moon is 60 times as far from the earth as an apple in a tree, so the force of earth's gravity on the moon is (A. 3,600 times weaker, B. 60 times stronger, C. 60 times weaker, D. the same).

T F 10. The law of gravity applies only to the sun, moon, and planets.

A B 11. If the moon were twice as far away, gravitational attraction between the earth and the moon would be (A. one-half B. one-fourth) as great.

12. Force of gravitational attraction between two objects is directly proportional to the product of their masses and inversely proportional to the square of the distance separating them.

T F 13. Scientists have proven that our sun is the only star that has planets orbiting it.

Answer T or F for true or false, fill in the blank, or select the letter for the phrase that best completes the sentence.

1. A simple machine changes the amount of force needed to do a job or the direction the force is applied. (same word)

A B C D 2. The Greek who said, "Give me a place to stand and a long enough lever, and I can move the world" was (A. Archimedes, B. Aristotle, C. Eratosthenes, D. Ptolemy).

A B 3. Mechanical advantage is found by dividing load by (A. effort, B. gravity).

A B C D 4. The tab on a soft drink can is an example of (A. an inclined plane, B. a lever, C. a pulley, D. a wheel and axle).

T F 5. The pivot point (fulcrum) of a lever must be located in the middle.

A B 6. If a load is moved closer to the fulcrum than the effort, the effort required to move the load will be (A. increased, B. reduced).

7. The Grand Canyon is about one mile deep, and the most popular trail out of the canyon is nine miles long; the mechanical advantage of the trail is 9.

A B C D 8. A screwdriver is an example of (A. a pulley, B. a ramp, C. a wheel and axle, D. an inclined plane).

T F 9. A screw is an inclined plane wrapped around a cylinder.

T F 10. Because of friction, the work produced by a simple machine is greater than the work put into a simple machine.

A B 11. The one that is likely to be the least efficient is (A. a simple machine, B. an 18-wheeler truck).

A B 12. A machine with no friction or other hindrance to its movement would have an efficiency of (A. zero, B. 100) percent.

Answer T or F for true or false, fill in the blank, or select the letter for the phrase that best completes the sentence.

T F 1. Energy is a term that has been in use for more than 2,000 years.

A B 2. Heat and light are examples of (A. matter, B. energy).

T F 3. Energy can be changed from one form to another.

A B C D 4. The equation E = f × d is used to find (A. efficiency, B. mechanical advantage, C. momentum, D. work).

5. Work transfers energy from one place to another.

A B C D 6. Foot-pounds (English system) and joules (metric system) both measure (A. force, B. mass, C. power, D. work).

7. James Prescott Joule found how mechanical energy due to motion compares to heat energy.

T F 8. Pushing against a desk that does not move is an example of work.

A B 9. The quantity that measures how quickly energy is supplied is called (A. work, B. power).

A B C 10. The English and metric system units for measuring power are (A. calorie and joule, B. pound and newton, C. horsepower and watt).

A B 11. The energy of motion is (A. kinetic, B. potential) energy.

T F 12. Doubling mass of a moving object doubles its kinetic energy.

T F 13. Doubling velocity of a moving object doubles its kinetic energy.

A B 14. An object would gain more kinetic energy by (A. doubling its mass, B. doubling its velocity).

A B 15. Stored energy is called (A. kinetic, B. potential) energy.

A B C 16. Almost every time that energy changes form, the amount of (A. heat, B. kinetic, C. potential) energy increases.

Answer T or F for true or false, fill in the blank, or select the letter for the phrase that best completes the sentence.

A B C D 1. Heat is a type of (A. energy, B. force, C. matter, D. temperature).

2. The three factors that determine the heat contained in an object are type of substance, mass, and temperature.

A B 3. The one that stores heat better is (A. iron, B. water).

A B 4. A thermometer works on the principle that most substances (A. contract, B. expand) when heated.

A B C D 5. The two most common substances used in thermometers are colored alcohol and (A. cooking oil, B. ethylene glycol, C. mercury, D. molten salt).

T F 6. Scientists are unable to measure temperatures greater than 1,700°F.

A B C D 7. The scientist who discovered that pure water has a fixed boiling and freezing temperature was (A. Anders Celsius, B. Antoine Lavoisier, C. Daniel Fahrenheit, D. John Dalton).

A B 8. High air pressure causes water to boil at a (A. higher, B. lower) temperature.

T F 9. Heat is the motion of atoms and molecules.

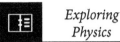

Answer T or F for true or false, or select the letter for the phrase that best completes the sentence.

(A) B — 10. Heat is a form of (A. kinetic, B. potential) energy.

(A) B C — 11. Heat moving from one end of a metal fireplace poker to the other end is an example of heat transfer by (A. conduction, B. convection, C. radiation).

(A) B — 12. The one that conducts heat better is (A. copper, B. wood).

(A) B — 13. Fur, feathers, and other substances with air pockets conduct heat (A. poorly, B. well).

A (B) C — 14. A sea breeze is set in motion because of (A. conduction, B. convection, C. radiation).

(T) F — 15. Heat is transferred from the sun to earth by radiation.

(T) F — 16. A steam engine works because heat flows from a hot region to a cold region.

A (B) — 17. A heat engine works best when the temperature change from heat source to heat sink is (A. about the same, B. greatly different).

(T) F — 18. Moving heat energy in a direction opposite to its normal flow requires work.

For More Study

19. The maximum efficiency possible for a machine that produces energy from the difference of ocean water at 18°C at the surface and 1°C at depth is 0.058.

Answer T or F for true or false, fill in the blank, or select the letter for the phrase that best completes the sentence.

T F 1. A rubber band is elastic because it will stretch.

T F 2. Steel is highly elastic.

3. The amount a solid object bends is directly proportional to the force acting on it.

A B 4. Spreading the weight of a solid over greater area (A. increases, B. reduces) pressure.

A B 5. The factor most important in producing water pressure is the (A. height, B. volume) of the water tank.

T F 6. The pressure of a liquid acts equally in all directions.

A B C D 7. Density is equal to mass divided by (A. area, B. pressure, C. volume, D. weight).

A B 8. As a hurricane approaches, air pressure will (A. increase, B. decrease).

A B C 9. The rate of diffusion of a gas is inversely proportional to the (A. square, B. square root, C. sum) of its molecular weight.

Matching

10. B Archimedes' principle of buoyancy

11. D Boyle's law

12. A Ideal gas law

13. C Bernoulli's principle

a. Pressure times volume of any gas divided by the temperature is a constant.

b. The lifting force acting on a solid object immersed in water is equal to the weight of the water shoved aside by the object.

c. The velocity of a fluid and its pressure are inversely related.

d. The volume of a gas is inversely proportional to the pressure.

Answer T or F for true or false, fill in the blank, or select the letter for the phrase that best completes the sentence.

T F 1. Waves are an efficient way to send energy from one place to another.

A B C D 2. The distance along a wave, including crest and trough, is its (A. axis, B. frequency, C. velocity, D. wavelength).

A B 3. The number of waves produced per time interval is its (A. frequency, B. velocity.)

A B C 4. Dividing how far a wave travels by the time it takes to travel that distance gives the wave's (A. frequency, B. velocity, C. wavelength).

5. The speed of any wave can be found by multiplying its frequency by its wavelength.

T F 6. Sound is produced by back and forth motion.

A B C D 7. The frequency of a sound is known as its (A. amplitude, B. color, C. pitch, D. velocity).

A B 8. The maximum displacement of a wave from its position of rest is its (A. amplitude, B. wavelength).

A B 9. A loud sound has (A. high, B. low) amplitude.

T F 10. A loud sound travels faster than a soft sound.

Answer T or F for true or false, fill in the blank, or select the letter for the phrase that best completes the sentence.

11. The pitch of a string on a stringed instrument depends on the length, thickness, and tension of the string.

A B 12. The (A. highest, B. lowest) pitch an object can make is known as its natural or fundamental frequency.

A B C D 13. The study of sound is known as (A. acoustics, B. astronomy, C. mechanics, D. thermodynamics).

A B 14. High frequency, ultrasonic sounds reflect (A. better, B. worse) from small objects than low frequency sounds.

A B 15. We perceive a light 100 times brighter as (A. 100 times, B. twice) as bright.

A B C D 16. The loudness of sound is measured in (A. candles, B. decibels, C. joules, D. watts).

T F 17. As sound waves spread out, they grow weaker by the square of the distance.

18. The three properties of a sound are frequency, intensity, and quality.

A B 19. Sound waves travel at the (A. same speed, B. different speeds) in air depending on its source.

A B 20. If pitch increases, then source and observer must be moving (A. toward, B. away from) one another.

Matching

1.__B__ brings light to a focus.

2.__C__ controls the amount of light that enters the eye.

3.__F__ is the opening through which light enters the eye.

4.__D__ adjusts light to the best focus.

5.__G__ is a surface of light sensitive nerves.

6.__E__ carries information from the eye to brain.

7.__H__ is sensitive to light but cannot see color.

8.__A__ is sensitive to light and can distinguish color.

a. Cones

b. Cornea

c. Iris

d. Lens

e. Optic nerve

f. Pupil

g. Retina

h. Rods

Answer T or F for true or false, fill in the blank, or select the letter for the phrase that best completes the sentence.

T F 9. Sunlight is a mixture of all the colors of the rainbow.

10. The eye has cones that can detect red, green, and blue________ light.

A B 11. The observation that light bounces from a mirror at the same angle at which it enters is known as the law of (A. reflection, B. refraction).

A B 12. The image behind a flat mirror is a (A. real, B. virtual) image.

A B 13. Most modern large telescopes use a (A. lens, B. mirror) to collect light and bring it to a focus.

A B 14. A lens thicker in the middle than at the edges is (A. convex, B. concave).

A B 15. The speed of light is (A. faster, B. slower) in water than in air.

A B 16. The bending of the sun's rays at sunset is an example of (A. refraction, B. reflection).

T F 17. The frequency of light is its brightness.

Answer T or F for true or false, fill in the blank, or select the letter for the phrase that best completes the sentence.

T F 1. Thales of Melitus discovered that amber could be given a charge of static electricity.

A B 2. The one that moves more freely is the (A. electron, B. proton).

T F 3. An object with a positive charge has more protons than electrons.

4. Coulomb's law of static electric force is very similar to Newton's law of gravity, but with charge replacing mass.

A B 5. The stronger force is (A. electrostatic, B. gravity).

A B 6. Glass is an example of a (A. conductor, B. nonconductor).

T F 7. All metals conduct electricity.

T F 8. No practical use has been found for battery-powered vehicles.

A B C D 9. The force that pushes electrons around a circuit is (A. resistance, B. charge, C. current, D. voltage).

10. State Ohm's law: current is directly proportional to voltage and inversely proportional to resistance

A B C D 11. The ohm is a unit for measuring (A. current, B. power, C. resistance, D. voltage).

12. To reduce the heating effect of electricity in wires, the current is reduced but the voltage is increased.

Answer T or F for true or false, or select the letter for the phrase that best completes the sentence.

T F 1. Unlike static electricity, magnetism was well understood from the time of the Greeks.

T F 2. William Gilbert proved that the compass is drawn toward the North Star.

T F 3. Earth's geographic North Pole and magnetic north pole have the same location.

A B C 4. The force that can attract but not repel is (A. electric charge, B. gravity, C. magnetism).

T F 5. A negative electric charge can be isolated without a corresponding positive electric charge.

T F 6. Magnets always have a north pole and a south pole.

A B C 7. The metal that a magnet attracts best is (A. aluminum, B. gold, C. iron).

A B C 8. Like magnetic poles (A. attract, B. repel) one another.

T F 9. The inverse square law states that a quantity decreases by the square of the distance.

T F 10. Gravity, static electricity, and magnetism all follow the inverse square law.

A B 11. The one that is more difficult to magnetize is (A. soft iron, B. steel).

A B 12. When magnetic domains become jumbled, magnetism is (A. lost, B. strengthened).

T F 13. An electric current can produce a magnetic field.

A B C D 14. The advantage of an electromagnet is that it (A. can be turned on and off, B. does not follow the inverse square law, C. can both attract and repel iron, D. takes less electricity to operate than a natural magnet).

T F 15. Michael Faraday discovered that a moving magnetic field generates electricity.

A B C D 16. Faraday succeeded in showing a connection between (A. chemistry and electricity, B. electricity and magnetism, C. magnetism and light, D. all of the above).

Answer T or F for true or false, fill in the blank, or select the letter for the phrase that best completes the sentence.

A B C D 1. The scientist who developed four equations that summarized electromagnetism was (A. Albert Einstein, B. Isaac Newton, C. James Clerk Maxwell, D. Michael Faraday).

A B C 2. The speed of electromagnetic waves is (A. greater, B. less, C. the same) as visible light.

A B C D 3. The first scientist to generate electromagnetic waves was (A. Arthur Compton, B. Guglielmo Marconi, C. Michael Faraday, D. Rudolf Hertz).

T F 4. FM radio waves carry around the world because they reflect from a layer in the upper atmosphere.

A B 5. The (A. AM, B. FM) radio band is prone to electrical interference.

A B C D 6. The M in AM and FM stands for (A. Marconi, B. Maxwell, C. modulation, D. momentum).

7. Write the numbers 1 to 4 in the blanks to rank these waves in order from lowest frequency (longest wavelength) to highest frequency (shortest wavelength): 3 blue visible light, 1 AM radio waves, 4 x rays, 2 infrared light.

A B C D 8. The period 1895–1905 is known as (A. the Aristotle period, B. the atomic age, C. the first scientific revolution, D. the second scientific revolution).

T F 9. In the photoelectric effect, the speed of electrons emitted depends on the brightness of the light.

T F 10. Albert Einstein explained the photoelectric effect by thinking of light as particles rather than waves.

A B C D 11. Albert Einstein won the Nobel Prize in physics because of his research papers about (A. Brownian motion, B. the equation $E = mc^2$, C. photoelectric effect, D. the special theory of relativity).

T F 12. Momentum is the product of mass and color.

T F 13. Arthur Compton found that an x ray could change the momentum of an electron.

T F 14. To date, the photoelectric effect is the only example of light acting as particles.

Answer T or F for true or false, or select the letter for the phrase that best completes the sentence.

A B C D 1. The one that is not made of smaller particles is (A. an electron, B. a hydrogen atom, C. a neutron, D. a proton).

A B C 2. The more massive subatomic particle is the (A. electron, B. neutron, C. proton).

A B C 3. A hydrogen atom changes to a helium atom if it gains (A. a neutron, B. a proton, C. an electron).

A B 4. The splitting of an atom into smaller parts is nuclear (A. fission, B. fusion).

T F 5. It is easier to cause the nuclei of atoms to break apart than to melt them together.

A B C D 6. Enrico Fermi discovered that the best particles to cause the break-up of a nucleus were (A. electrons, B. helium nuclei, C. neutrons, D. protons).

A B C D 7. The first person to state that uranium could undergo fission and produce a self-sustaining chain reaction was (A. Albert Einstein, B. Enrico Fermi, C. Franklin D. Roosevelt, D. Lise Meiter).

T F 8. A breeder reactor changes uranium-238 into plutonium.

A B C D 9. The purpose of a moderator is to (A. absorb neutrons, B. cool the reactor, C. generate heat, D. slow neutrons).

A B 10. The total mass after a nuclear reaction is (A. less, B. more) than the total mass before the reaction.

T F 11. More than half of the electricity used in the United States comes from nuclear reactors.

T F 12. The first cold fusion reactor was built in 2002.

Answer T or F for true or false, fill in the blank, or select the letter for the phrase that best completes the sentence.

A B 1. In the black box experiment, the amount of ultraviolet light was (A. far less, B. many times greater) than predicted.

T F 2. Max Planck found that the smallest quantum of light was proportional to its speed.

A B 3. An ultraviolet quantum has (A. more, B. less) energy than an infrared quantum.

T F 4. Electrons change orbits only by absorbing or emitting set amounts of energy.

A B 5. The one with a wave long enough to be detected is (A. an electron, B. a baseball).

T F 6. Light waves can interfere with one another.

A B 7. The ground state, or lowest orbit of an electron, corresponds to its (A. fundamental frequency, B. highest overtone).

A B C D 8. The first particle shown to have a wave nature was (A. an alpha particle, B. an electron, C. a gamma ray, D. a proton).

T F 9. The properties that we observe about an electron depend on the experiment that we devise to study the electron.

10. The Heisenberg uncertainty principle states that the precise position, mass, and velocity for any particle cannot be determined exactly.

Matching

11. A Niels Bohr

12. D Louis de Borglie

13. B Max Planck

14. C Werner Heisenberg

a. Developed a model of the atom and electron orbits.

b. Explained black body radiation by using energy quanta.

c. Developed the uncertainty principle.

d. Proposed matter waves and calculated their wavelengths.

Quizzes and Tests Section

Q	*Exploring Mathematics* Concepts & Comprehension	Quiz 1	Scope: Chapters 1–4	Total score: ____of 100	Name: Date:

Matching (2 Points Each Question)

1. C day — a. due to the tilt of the earth's axis, equal to three months
2. E week — b. earth revolves around the sun once
3. D month — c. earth rotates on its axis once
4. A season — d. moon revolves around the earth once
5. B year — e. seven days

Fill-in-the-Blank Questions (4 Points Each Question)

6. The length of a mile in feet is 5,280.
7. "A pint is a pound the world around."

Multiple Choice Questions (4 Points Each Question)

8. The first calendar with a leap day every four years was the one
   A. authorized by Julius Caesar B. used by the American colonies after 1752
   C. used by the Babylonians D. used by the Egyptians
9. The inventors of the hourglass were the
   A. Babylonians B. British Navy
   C. Egyptians D. Romans
10. Military time has hours numbered from 0000 to
   A. 0400 B. 1200
   C. 2400 D. 3600
11. Time zones were introduced when it became common to travel by
   A. airplanes B. ox carts
   C. ships D. trains
12. NASA's Climate Orbiter to Mars failed because
   A. American and French engineers did not communicate with one another
   B. engineers used two different measures of force
   C. fuel had been measured improperly
   D. the spacecraft weighed too much
13. A scruple was a standard of weight for measuring
   A. barley B. diamonds
   C. drugs D. potatoes
14. Most early measures of distance were based on
   A. animal strides B. the human body
   C. parts of ships D. Roman military terms

15. Currently, the meter is defined as

A. 1,640,763.73 wavelengths of krypton gas
B. 1/10,000,000 of the distance from the equator to the North Pole
C. the distance between two scratch marks on a metal rod
D. the distance light travels in 1/299,792,458 of a second

16. Daniel Fahrenheit set the boiling temperature of water on his thermometer at

A. 0 degrees B. 32 degrees
C. 100 degrees D. 212 degrees

Multiple Answer Questions (3 Points Each Answer)

17. Using the Babylonian calendar of 360 days in a year, how many days are in one-third of a year; one-fifth of a year; one-twentieth of a year; one-sixtieth of a year?

a. 120 days b. 72 days c. 18 days d. 6 days

18. A hand is four inches. How tall is a horse in inches that is 15 hands tall? How tall in feet?

a. 60 inches b. 5 feet

Short Answer Questions (4 Points Each Question)

19. What is the main reason to have leap days?
to mach seasons with the calendor

20. Assume that the first four-hour watch began at midnight. What time would it be at five bells on the second watch? 6:30 am

21. At 4:00 p.m., a family on vacation drives from Mountain Standard Time into Central Standard Time. Should their watches be set one hour earlier to 3:00 p.m. or one hour later to 5:00 p.m.?
hour later

22. The tallest mountain on earth is Mt. Everest. Its summit is 29,035 feet above sea level. How high is the mountain in miles? 5.499

Applied Learning Activities (2 Points Each Answer)

23. Feel your pulse at the wrist and count the number of beats in a minute. Calculate the number of times your heart beats in a day. 100,000 ish

Choose the larger:

24. A. foot B. yard
25. A. fathom B. yard
26. A. nautical mile B. statute mile
27. A. cup B. quart
28. A. bushel B. peck

Q	*Exploring Mathematics* Concepts & Comprehension	Quiz 2	Scope: Chapters 5–8	Total score: ____of 100	Name: Date:

Matching (2 Points Each Question)

1. __E__ circle
2. __A__ pentagon
3. __D__ rectangle
4. __C__ right triangle
5. __B__ square

a. a polygon with five sides
b. a rectangle with four equal sides
c. a polygon with three sides and one right angle
d. a quadrilateral with opposite sides parallel and equal in length
e. is not a polygon

6. __D__ 2L + 2W
7. __E__ 4S
8. __F__ Ah
9. __B__ L x W
10. __A__ pr2
11. __C__ S2

a. area of a circle
b. area of a rectangle
c. area of a square
d. perimeter of a rectangle
e. perimeter of a square
f. volume

12. __E__ Archimedes
13. __A__ Eucli
14. __C__ Johannes Kepler
15. __D__ Pythagoras
16. __B__ Thales

a. discovered that the sum of the 3 angles of any triangle is 180 degrees
b. used ratios to find the heights of buildings
c. proved planets follow elliptical orbits
d. wrote *Elements of Geometry*
e. ancient Greek who worked out a way to show large numbers that he called myriads

17. __B__ circle
18. __D__ ellipse
19. __A__ parabola
20. __C__ hyperbola

a. a mirror of this shape will focus sunlight
b. all points are the same distance from the center
c. the first part of the name means over or beyond
d. the orbit of Halley's comet is of this shape

21. __B__ 1881, 121, 1001
22. __C__ 2, 3, 5, 7, 11, 13 . . .
23. __A__ 1, 1, 2, 3, 5, 8, 13 . . .
24. __D__ 1, 4, 9, 16, 25 . . .
25. __E__ 1, 3, 6, 10, 15 . . .

a. Fibonacci numbers
b. palindromes
c. prime numbers
d. square numbers
e. triangular numbers

Fill-in-the-Blank Questions (2 Points Each Answer)

26. The sum of the squares of the legs of a right triangle are equal to the Square of the hypotenuse.

27. The next Fibonacci number after 89 and 144 is 233.

Multiple Choice Questions (4 Points Each)

28. The Egyptian knotted rope was used to measure out
    A. a pyramid with sloping sides B. a rectangle with parallel sides
    C. a silo of a fixed height D. a triangle with a right angle

29. Doubling the length, width, and height of a box gives it a volume
    A. twice as great B. three times as great
    C. six times as great D. eight times as great

30. A whispering gallery has a shape like
    A. a circle B. a hyperbola C. a parabola D. an ellipse

31. The study of the properties of whole numbers is called
    A. algebra B. geometry C. number theory D. real analysis

True and False (2 Points Each)

32. T F Mathematics is sometimes called the ruler of science.

33. T F Isaac Newton introduced the use of place value and the numeral 0 to Europe.

34. T F The prefix *bi* means one-half.

35. T F The word billion has the same meaning in England as in the United States.

36. T F Of the prefixes giga, mega, and tera, the one that has the greatest value is mega.

Short Answer Questions (4 Points)

37. A room is 10 feet wide and 14 feet long. How many square tiles, one foot on a side, would be needed to completely cover the room? 140 tiles

Applied Learning Activity (14 Points Total: 2 Points each Square)

38. Draw a Fibonacci spiral.

Q	*Exploring Mathematics* Concepts & Comprehension	Quiz 3	Scope: Chapters 9–11	Total score: ____of 100	Name: Date:

Matching (2 Points Each Question)

1. A 1, 2, 3, 4, 5 — a. counting numbers
2. G 0, 1, 2, 3, 4, 5 — b. even numbers
3. B 2, 4, 6, 8, 10 — c. integers
4. E 1, 3, 5, 7, 9 — d. irrational numbers
5. C -3, -2, -1, 0, +1, +2, +3 — e. odd numbers
6. F 1/1, 1/2, 3/4, 2/3 — f. rational numbers
7. D SR2, p, (1 + SR5)/2 — g. whole numbers

Use the figure to match the equations (questions 8–11) and the statements (questions 12–15)

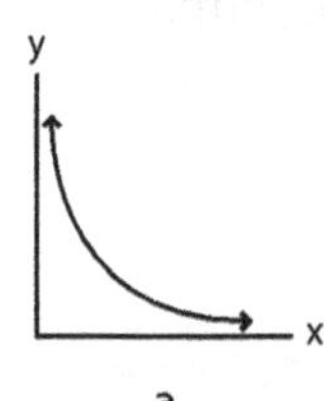

a

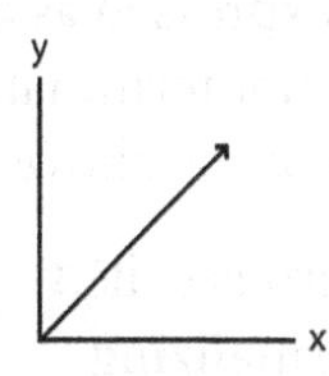

b

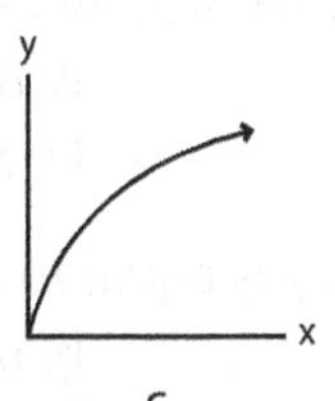

c

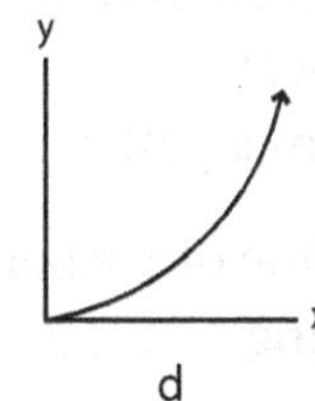

d

Equations:

8. B $y = kx$ 9. D $y = kx2$ 10. C $y = kSRx$ 11. A $y = k / x$

Statements:

12. B The length a spring stretches (y-axis) is directly proportional to the force pulling on the spring (x-axis).
13. D The distance an object falls (y-axis) under the influence of gravity is directly proportional to the square of the time it has fallen (x-axis).
14. A The volume of a gas (y-axis) is inversely proportional to the pressure acting on the gas (x-axis).
15. C The period of a pendulum (y-axis) is directly proportional to the square root of the length of the pendulum (x-axis).

16. C Andrew Wiles — a. discovered how to calculate the coefficients of a binomial raised to a power
17. B Blaise Pascal — b. he called his triangle an arithmetic triangle
18. A Isaac Newton — c. his last theorem was solved in 1995
19. D Leonhard Euler — d. solved the Konigsberg bridge problem
20. E Pierre de Fermat — e. proved that xn + yn = zn has no solution with whole numbers except for n = 2

Multiple Choice Questions (4 Points Each Question)

21. Before the invention of calculators, shares were used to reduce the necessity of doing
    A. addition B. subtraction
    C. multiplication D. division

22. The American colonies divided the real, a Spanish coin, into how many pieces?
    A. 2 B. 4 C. 8 D. 12

23. Two bits is equal to how many cents?
    A. 12½ B. 25 C. 50 D. 100

24. A common fraction can be changed into a decimal by dividing the numerator by the
    A. denominator B. greatest common factor
    C. least common multiple D. remainder

25. The square root of two, is an example of
    A. a common fraction B. an irrational number
    C. a rational number D. a terminating decimal

26. The digits of the square root of two, $\sqrt{2}$, when expressed as a decimal
    A. do not repeat B. do not terminate
    C. do not form a pattern D. all of the above

27. Discovering the value of x when y is equal to zero is called
    A. modernizing B. normalizing
    C. solving D. zeroing the equation

28. The 5 in the equation $5x + 2 = 0$ is called
    A. a coefficient B. a constant
    C. an equalizer D. a variable

29. In the 1600s, the French mathematician Rene Descartes discovered a way to combine algebra with
    A. computer programming B. geometry
    C. number theory D. physics

30. The problem that a computer helped solve was the
    A. bell peal problem B. binomial theorem
    C. four-color map problem D. Konigsberg bridge problem

31. The expression 3! is read as "three
    A. combinations" B. factorial"
    C. permutations" D. probabilities"

32. The value of 5! is
    A. 24 B. 25 C. 120 D. 125

Applied Learning Activities (4 Points)

33. The state of Missouri has license plates with three letters followed by three digits. How many license plates are possible? 17,576,000

(8 Points Total; 2 Points Each Answer)

34. Four Corners is the only point in the United States where four states touch a single point. Name the states.

utah, calorado, arazona, new mexico

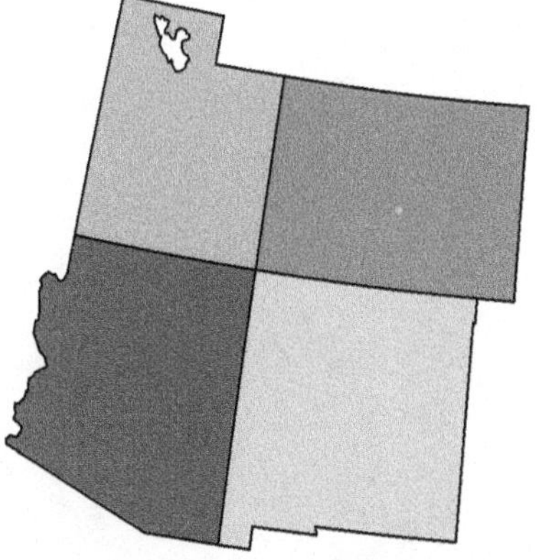

Matching (2 Points Each Question)

1. ____ Augusta Ada Byron, Lady Lovelace
2. ____ Howard H. Aiken
3. ____ Charles Babbage
4. ____ Herman Hollerith
5. ____ Johannes Kepler
6. ____ Gottfried Leibnitz
7. ____ John Napier
8. ____ Blaise Pascal

a. built a calculator called the Step Reckoner
b. built a calculator to help his father, a tax collector
c. built the first general-purpose calculating machine
d. built the difference engine
e. invented logarithms
f. invented tabulating machines used in 1890 census
g. spent six years calculating the orbit of Mars
h. wrote the first computer program

Fill-in-the-Blank Questions (2 Points Each Answer)

9. The expression log103 = 0.477 is read as "The logarithm of the number __________ in base __________ is __________."

10. Base 10 uses the digits 0 through ______________.

Multiple Choice Questions (5 Points Each Question)

11. Any number raised to the zero power is
A. 0 B. 1 C. 2 D. undefined

12. The number 5,280 changed to standard notation is
A. .5280 x 101 B. 5,280 x 103 C. 5.28 x 102 D. 5.28 x 103

13. In the early days of computers, input was mainly by
A. colored ribbons B. punched cards
C. spoken commands D. switches and relays

14. The letters RAM stand for
A. random access memory B. reasonably accurate member
C. recent abacus modification D. Robert A. Morley

15. In computer usage, a single position for a binary digit is called
A. bit B. byte C. kilo D. pixel

16. In personal computers, a byte of data is made of how many bits?
A. one B. two C. eight D. ten

17. In 1861, James Clerk Maxwell made a color photograph using
A. computer enhancement B. color ink drops sprayed on paper
C. polarized light D. the three colors of red, green, and blue

18. Video images can be compressed by
    A. converting black and white images to color images
    B. having reporters avoid standing in front of a blue sky
    C. transmitting all pixels that are the same as the previous one
    D. transmitting only those pixels that are different from the previous one

19. Moore's law states that computers double in power every 18
    A. days   B. months   C. decades   D. years

20. A computer with components put farther apart will run more slowly because
    A. electric signals can go no faster than the speed of light
    B. larger components must be made of less costly materials
    C. of resistance in the wires
    D. the electrons get lost

Short Answer Questions (4 Points Each Question)

21. In this list, which one is considered the "heart" of a computer: input, control program, memory, central processing unit, output? Centra Procesing unit

22. The Constitution of the United States has 4,609 words and 26,747 characters. At the rate of 7,000 bytes per second, how long would it take a computer to download the Constitution of the United States as an uncompressed text file? 3.8

Applied Learning Activities (6 Points Each Puzzle/Riddle)

23. Samson's Riddle — The Bible has puzzles such as Samson's riddle in Judges 14:14: He replied, "Out of the eater, something to eat; out of the strong, something sweet." Hint: You can find the answer in Judges 14:8.

24. On the road to St. Ives — Try to solve this people-on-the-road puzzle that was turned into an English children's rhyme:

    As I was going to St. Ives I met a man with seven wives;
    Every wife had seven sacks; Every sack had seven cats;
    Every cat had seven kits [kittens]; Kits, cats, sacks, and wives,
    How many were going to St. Ives?

    Can you figure out the answer to the riddle?

25. River Crossing — A canoeist must cross a river with three things, but his canoe can hold only one thing at a time. How can the canoeist get a wolf, goat, and carrots across a river? If left alone, the wolf would eat the goat, and the goat would eat the carrots.

T	*Exploring Mathematics* Concepts & Comprehension	Test	Scope: Chapters 1–14	Total score: ____of 100	Name: Date:

Matching (1 Points Each Question)

1. __C__ day	a. due to the tilt of the earth's axis, equal to three months
2. __E__ week	b. earth revolves around the sun once
3. __D__ month	c. earth rotates on its axis once
4. __A__ season	d. moon revolves around the earth once
5. __B__ year	e. seven days

6. __B__ circle	a. a mirror of this shape will focus sunlight
7. __D__ ellipse	b. all points are the same distance from the center
8. __A__ parabola	c. the first part of the name means over or beyond
9. __C__ hyperbola	d. the orbit of Halley's comet is of this shape

10. __A__ 1, 2, 3, 4, 5	a. counting numbers
11. __G__ 0, 1, 2, 3, 4, 5	b. even numbers
12. __B__ 2, 4, 6, 8, 10	c. integers
13. __E__ 1, 3, 5, 7, 9	d. irrational numbers
14. __C__ -3, -2, -1, 0, +1, +2, +3	e. odd numbers
15. __F__ 1/1, 1/2, 3/4, 2/3	f. rational numbers
16. __D__ SR2, p, (1 + SR5)/2	g. whole numbers

17. __H__ Augusta Ada Byron, Lady Lovelace	a. built a calculator called the Step Reckoner
18. __D__ Howard H. Aiken	b. built a calculator to help his father, a tax collector
19. __C__ Charles Babbage	c. built the first general-purpose calculating machine
20. __F__ Herman Hollerith	d. built the difference engine
21. __G__ Johannes Kepler	e. invented logarithms
22. __E__ Gottfried Leibnitz	f. invented tabulating machines used in 1890 census
23. __A__ John Napier	g. spent six years calculating the orbit of Mars
24. __B__ Blaise Pascal	h. wrote the first computer program

Fill-in-the-Blank Questions (2 Points Each Answer)

25. The length of a mile in feet is 5,280.

26. "A pint is a pound the world around."

27. The sum of the Squares of the legs of a right triangle are equal to the Square of the hypotenuse.

28. The next Fibonacci number after 89 and 144 is 233.

29. The expression log103 = 0.477 is read as "The logarithm of the number 3 in base 10 is 0.477."

30. Base 10 uses the digits 0 through 9.

Multiple Choice Questions (2 Points Each Question)

31. NASA's Climate Orbiter to Mars failed because
    A. American and French engineers did not communicate with one another
    B. engineers used two different measures of force
    C. fuel had been measured improperly
    D. the spacecraft weighed too much

32. A scruple was a standard of weight for measuring
    A. barley C. drugs B. diamonds D. potatoes

33. Currently, the meter is defined as
    A. 1,640,763.73 wavelengths of krypton gas
    B. 1/10,000,000 of the distance from the equator to the North Pole
    C. the distance between two scratch marks on a metal rod
    D. the distance light travels in 1/299,792,458 of a second

34. Daniel Fahrenheit set the boiling temperature of water on his thermometer at
    A. 0 degrees C. 100 degrees B. 32 degrees D. 212 degrees

35. The Egyptian knotted rope was used to measure out
    A. a pyramid with sloping sides B. a rectangle with parallel sides
    C. a silo of a fixed height D. a triangle with a right angle

36. Doubling the length, width, and height of a box gives it a volume
    A. twice as great B. three times as great
    C. six times as great D. eight times as great

37. A whispering gallery has a shape like
    A. a circle B. a hyperbola C. a parabola D. an ellipse

38. The study of the properties of whole numbers is called
    A. algebra B. geometry C. number theory D. real analysis

39. A common fraction can be changed into a decimal by dividing the numerator by the
    A. denominator B. greatest common factor
    C. least common multiple D. remainder

40. The square root of two, is an example of

A. a common fraction
B. an irrational number
C. a rational number
D. a terminating decimal

41. In the 1600s, the French mathematician Rene Descartes discovered a way to combine algebra with

A. computer programming
B. geometry
C. number theory
D. physics

42. The problem that a computer helped solve was the

A. bell peal problem
B. binomial theorem
C. four-color map problem
D. Konigsberg bridge problem

43. In 1861, James Clerk Maxwell made a color photograph using

A. computer enhancement
B. color ink drops sprayed on paper
C. polarized light
D. the three colors of red, green, and blue

44. Video images can be compressed by

A. converting black and white images to color images
B. having reporters avoid standing in front of a blue sky
C. transmitting all pixels that are the same as the previous one
D. transmitting only those pixels that are different from the previous one

45. Moore's law states that computers double in power every 18

A. days B. months C. decades D. years

46. A computer with components put farther apart will run more slowly because

A. electric signals can go no faster than the speed of light
B. larger components must be made of less costly materials
C. of resistance in the wires
D. the electrons get lost

Multiple Answer Question (3 Points Each Answer)

47. A hand is four inches. How tall is a horse in inches that is 15 hands tall? How tall in feet?

a. 60 inches b. 5 feet

Short Answer Questions (2 Points Each \Question)

48. What is the main reason to have leap days?

to keep seasons on track

49. At 4:00 p.m., a family on vacation drives from Mountain Standard Time into Central Standard Time. Should their watches be set one hour earlier to 3:00 p.m. or one hour later to 5:00 p.m.?

later

50. The tallest mountain on earth is Mt. Everest. Its summit is 29,035 feet above sea level. How high is the mountain in miles? 5.499

51. A room is 10 feet wide and 14 feet long. How many square tiles, one foot on a side, would be needed to completely cover the room? 140

Applied Learning Activities (6 Points Each Puzzle)

52. Sock Puzzle — Because of an electrical power failure, a boy must get dressed in a dark bedroom. His sock drawer has 10 blue socks and 10 black socks, but in the darkness he cannot tell them apart. He dresses anyway. He reaches into the drawer to grab spare socks so he can change into matching colors later. How many should he take to be certain he has a matching pair?

One

53. Durer's Number Square — Make a number square by using the digits 1 through 9 in a three by three square. Each of the rows, columns, and diagonals should add to the same number. Eight different squares are possible.

Q	*Exploring Physics* / Concepts & Comprehension	Quiz 1	Scope: Chapters 1–4	Total score: ____of 100	Name: / Date:

Matching (2 Points Each)

1. E first law of motion — a. $a = f/m$
2. A second law of motion — b. $f = m \times a$
3. C third law of motion — c. $f_{ab} = -f_{ba}$
4. B force equation — d. $I = f \times t$
5. D definition of impulse — e. If $f = 0$ then $a = 0$
6. F definition of momentum — f. $p = m \times v$

Fill-in-the-Blank Questions (4 Points Each Answer)

7. To calculate speed, divide distance by time.

8. Suppose a canoeist takes 70 days to paddle the entire length of the Mississippi River, a distance of 3,710 miles. The canoeist's average speed in miles per day is 53.

9. On the moon, the acceleration due to gravity is 5.3 ft/sec$_2$ rather than 32 ft/sec$_2$. If an object fell six seconds before hitting the ground, it strikes the ground with a speed of 31.8 ft/sec. (Hint: Use the final velocity equation.)

10. Momentum is the mass of an object times its velocity.

11. Force of gravitational attraction between two objects is directly proportional to the product of their masses and inversely proportional to the square of the distance separating them.

12. The Grand Canyon is about one mile deep, and the most popular trail out of the canyon is nine miles long; the mechanical advantage of the trail is 9.

Multiple Choice Questions (4 Points Each)

13. Physics is the science that explores how energy acts on
    A. heat, B. light, (C.) matter, D. sound.

14. A feather and lump of lead will fall at the same speed in
    A. a high speed wind tunnel, B. the atmosphere, (C.) a vacuum, D. water

15. To study the motion of falling objects, Galileo
    A. beat them into cubes
    B. dropped them from a high tower
    C. pushed them from a cliff
    (D.) rolled them down a ramp

16. Acceleration is found by dividing the
    A. average velocity
    B. distance
    C. gravity
    (D.) change in speed by the change in time.

17. Inertia is a property of matter that resists changing its
A. electric charge, B. mass, C. momentum, D. velocity

18. Kepler proved that planets traveled in orbits that were
A. circular, B. elliptical, C. parabolic, D. straight-line

19. The Greek who said, "Give me a place to stand and a long enough lever, and I can move the world" was
A. Archimedes, B. Aristotle, C. Eratosthenes, D. Ptolemy

20. The tab on a soft drink can is an example of
A. an inclined plane, B. a lever, C. a pulley, D. a wheel and axle

Short Answer Questions (4 Points Each Question)

21. An ordinary passenger car can accelerate to 60 miles per hour in about eight seconds. What is the car's acceleration? 7.5 mi/hr

22. State the second law of motion.

Applied Learning Activities (2 Points Each Answer – 20 Points Total)

23. Label the four forces acting on an airplane in flight.

24. Label the fulcrum, load, and effort points on the seesaw.

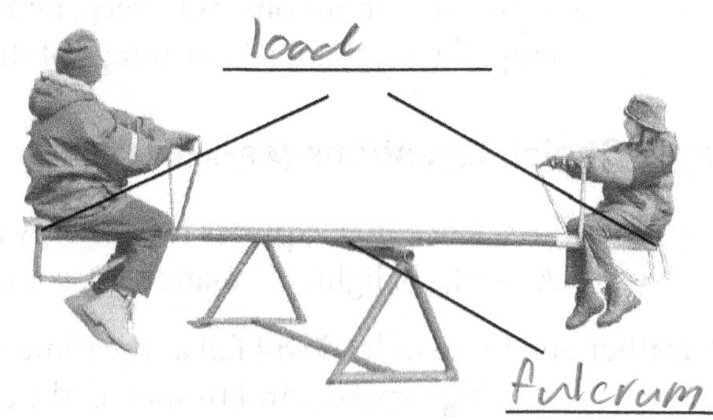

25. Label the fulcrum, load, and effort points on the nutcracker.

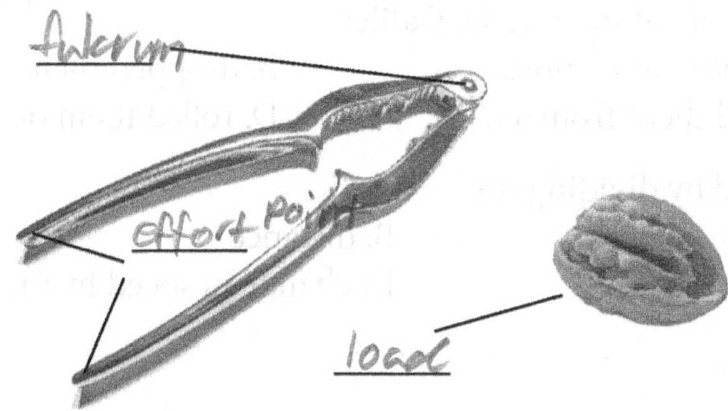

Q	*Exploring Physics* Concepts & Comprehension	Quiz 2	Scope: Chapters 5–7	Total score: ____of 100	Name: Date:

Matching (2 Points Each)

1. B Archimedes' principle of buoyancy
2. D Boyle's law
3. A Ideal gas law
4. C Bernoulli's principle

a. Pressure times volume of any gas divided by the temperature is a constant.

b. The lifting force acting on a solid object immersed in water is equal to the weight of the water shoved aside by the object.

c. The velocity of a fluid and its pressure are inversely related.

d. The volume of a gas is inversely proportional to the pressure.

Fill-in-the-Blank Questions (4 Points Each Answer)

5. Work transfers energy from one place to another.

6. James Prescott Joule found how mechanical energy due to motion compares to heat energy.

7. The three factors that determine the heat contained in an object are type of substance, mass, and temperature.

8. The maximum efficiency possible for a machine that produces energy from the difference of ocean water at 18°C at the surface and 1°C at depth is 058.

Multiple Choice Questions (4 Points Each)

9. The equation E = f x d is used to find
A. efficiency B. mechanical advantage
C. momentum (D) work

10. Foot-pounds (English system) and joules (metric system) both measure
A. force B. mass C. power (D) work

11. Heat is a type of
(A) energy B. force C. matter D. temperature

12. The two most common substances used in thermometers are colored alcohol and
A. cooking oil B. ethylene glycol (C) mercury D. molten salt

13. The scientist who discovered that pure water has a fixed boiling and freezing temperature was
A. Anders Celsius B. Antoine Lavoisier (C) Daniel Fahrenheit D. John Dalton

14. Density is equal to mass divided by
A. area B. pressure (C) volume D. weight

15. The English and metric system units for measuring power are
A. calorie and joule B. pound and newton C. horsepower and watt

16. Almost every time energy changes form, the amount of what kind of energy increases?
A. heat B. kinetic C. potential

17. Heat moving from one end of a metal fireplace poker to the other end is an example of heat transfer by
A. conduction B. convection C. radiation

18. A sea breeze is set in motion because of
A. conduction B. convection C. radiation

19. The rate of diffusion of a gas is inversely proportional to the ______ of its molecular weight.
A. square B. square root C. sum

True and False (2 Points Each)

20. T F Energy is a term that has been in use for more than 2,000 years.
21. T F Energy can be changed from one form to another.
22. T F Pushing against a desk that does not move is an example of work.
23. T F Doubling mass of a moving object doubles its kinetic energy.
24. T F Doubling velocity of a moving object doubles its kinetic energy.
25. T F Scientists are unable to measure temperatures greater than 1,700°F.
26. T F Heat is the motion of atoms and molecules.
27. T F Heat is transferred from the sun to earth by radiation.
28. T F A steam engine works because heat flows from a hot region to a cold region.
29. T F Moving heat energy in a direction opposite to its normal flow requires work.
30. T F A rubber band is elastic because it will stretch.
31. T F Steel is highly elastic.
32. T F The pressure of a liquid acts equally in all directions.

Applied Learning Activity (6 Points)

33. What does this demonstration measure? heat capacity

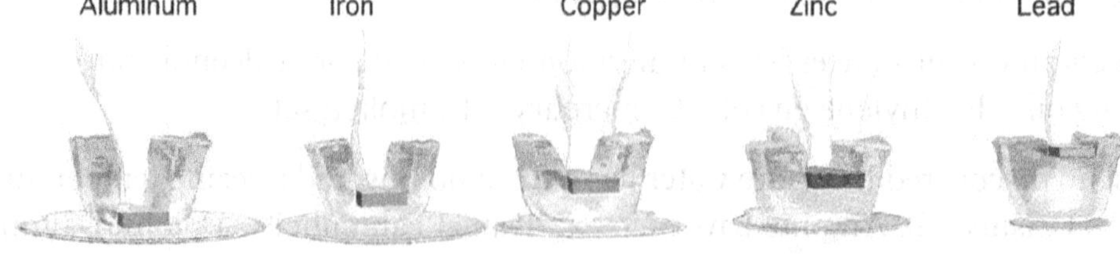

Q	*Exploring Physics* Concepts & Comprehension	Quiz 3	Scope: Chapters 8–10	Total score: ____of 100	Name: Date:

Matching (2 Points Each Question)

1. B brings light to a focus. — a. Cones
2. C controls the amount of light that enters the eye. — b. Cornea
3. F is the opening through which light enters the eye. — c. Iris
4. D adjusts light to the best focus. — d. Lens
5. G is a surface of light sensitive nerves. — e. Optic nerve
6. E carries information from the eye to the brain. — f. Pupil
7. H is sensitive to light but cannot see color. — g. Retina
8. A is sensitive to light and can distinguish color. — h. Rods

Fill-in-the-Blank Questions (2 Points Each Answer)

9. The pitch of a string on a stringed instrument depends on the length, thickness, and tension of the string.
10. The three properties of a sound are frequency, intensity, and quality.
11. The eye has cones that can detect red, green, and blue light.
12. Coulomb's law of static electric force is very similar to Newton's law of gravity, but with charge replacing mass.
13. To reduce the heating effect of electricity in wires, the current is reduced but the Voltage is increased.

Underline the Correct Answer (2 Points Each Answer)

14. The (A. highest, B. lowest) pitch an object can make is known as its natural or fundamental frequency.
15. High frequency, ultrasonic sounds reflect (A. better, B. worse) from small objects than low frequency sounds.
16. Sound waves travel at the (A. same speed, B. different speeds) in air depending on their source.
17. If pitch increases, then source and observer must be moving (A. toward, B. away from) one another.
18. The observation that light bounces from a mirror at the same angle at which it enters is known as the law of (A. reflection, B. refraction).
19. The image behind a flat mirror is a (A. real, B. virtual) image.
20. Most modern, large telescopes use a (A. lens, B. mirror) to collect light and bring it to a focus.
21. A lens thicker in the middle than at the edges is (A. convex, B. concave).

22. The speed of light is (A. faster, B. slower) in water than in air.

23. The bending of the sun's rays at sunset is an example of (A. refraction, B. reflection).

24. The one that moves more freely is the (A. electron, B. proton).

25. The stronger force is (A. electrostatic, B. gravity).

26. Glass is an example of a (A. conductor, B. nonconductor).

Multiple Choice Questions (2 Points Each Question)

27. The distance along a wave, including crest and trough, is its
    A. axis, B. frequency, C. velocity, D. wavelength

28. The frequency of a sound is known as its A. amplitude, B. color, C. pitch, D. velocity

29. The study of sound is known as A. acoustics, B. astronomy, C. mechanics, D. thermodynamics

30. The loudness of sound is measured in A. candles, B. decibels, C. joules, D. watts

31. The force that pushes electrons around a circuit is A. resistance, B. charge, C. current, D. voltage

32. The ohm is a unit for measuring A. current, B. power, C. resistance, D. voltage

Short Answer Question (6 Points)

33. State Ohm's law: current is proporshonal to its voltage

True and False (2 Points Each)

34. T F As sound waves spread out, they grow weaker by the square of the distance.

35. T F Sunlight is a mixture of all the colors of the rainbow.

36. T F The frequency of light is its brightness.

37. T F Thales of Melitus discovered that amber could be given a charge of static electricity.

38. T F An object with a positive charge has more protons than electrons.

39. T F All metals conduct electricity.

40. T F No practical use has been found for battery-powered vehicles.

Applied Learning Activity (2 Points)

Identify the sources for each decibel level:

Decibels

41. 8 G
42. 10–20 D
43. 20–30 F
44. 40–50 E
45. 50–60 B
46. 70–80 A
47. 90–100 H
48. 110 C

a. Heavy street traffic

b. Automobile

c. Thunder

d. Whisper

e. Ordinary conversation

f. Average home sounds (such as the humming of a refrigerator)

g. Rustling of leaves

h. Jack hammer

Matching (3 Points Each Question)

1. __A__ Niels Bohr
2. __D__ Louis de Borglie
3. __B__ Max Planck
4. __C__ Werner Heisenberg

a. developed a model of the atom and electron orbits
b. explained black body radiation by using energy quanta
c. developed the uncertainty principle
d. proposed matter waves and calculated their wavelengths

Fill-in-the-Blank Question (4 Points)

5. The Heisenberg uncertainty principle states that the precise position, mass, and __velocity__ cannot be determined exactly.

Multiple Choice Questions (4 Points Each Question)

6. The advantage of an electromagnet is that it
   (A.) can be turned on and off   B. does not follow the inverse square law
   C. can both attract and repel iron   D. takes less electricity to operate than a natural magnet

7. Faraday succeeded in showing a connection between
   A. chemistry and electricity   B. electricity and magnetism
   C. magnetism and light   (D.) all of the above

8. The scientist who developed four equations that summarized electromagnetism was
   A. Albert Einstein   B. Isaac Newton
   (C.) James Clerk Maxwell   D. Michael Faraday

9. The first scientist to generate electromagnetic waves was
   A. Arthur Compton   B. Guglielmo Marconi
   C. Michael Faraday   (D.) Rudolf Hertz

10. The M in AM and FM stands for
    A. Marconi, B. Maxwell, (C.) modulation, D. momentum

11. The period 1895–1905 is known as
    A. the Aristotle period   B. the atomic age
    (C.) the first scientific revolution   D. the second scientific revolution

12. Albert Einstein won the Nobel Prize in physics because of his research papers about
    A. Brownian motion   B. the equation E = mc2
    (C.) photoelectric effect   D. special theory of relativity

13. The one that is not made of smaller particles is
    (A.) an electron   B. a hydrogen atom   C. a neutron   D. a proton

14. Enrico Fermi discovered that the best particles to cause the break-up of a nucleus were
    A. electrons   B. helium nuclei   (C.) neutrons   D. protons

15. The first person to state that uranium could undergo fission and produce a self-sustaining chain reaction was
A. Albert Einstein B. Enrico Fermi C. Franklin D. Roosevelt D. Lise Meiter

16. The purpose of a moderator is to
A. absorb neutrons B. cool the reactor C. generate heat D. slow neutrons

17. The first particle shown to have a wave nature was
A. an alpha particle B. an electron C. a gamma ray D. a proton

Underline the Correct Answer (2 Points Each Answer)

18. The one that is more difficult to magnetize is (A. soft iron, B. steel).

19. When magnetic domains become jumbled, magnetism is (A. lost, B. strengthened).

20. The (A. AM, B. FM) radio band is prone to electrical interference.

21. The splitting of an atom into smaller parts is nuclear (A. fission, B. fusion).

22. The total mass after a nuclear reaction is (A. less, B. more) than the total mass before the reaction.

23. In the black box experiment, the amount of ultraviolet light was (A. far less, B. many times greater) than predicted.

24. An ultraviolet quantum has (A. more, B. less) energy than an infrared quantum.

25. The one with a wave long enough to be detected is (A. an electron, B. a baseball).

26. The ground state, or lowest orbit of an electron, corresponds to its (A. fundamental frequency, B. highest overtone).

Applied Learning Activities (2 Points Each Answer)

27. Write the numbers 1 to 4 in the blanks to rank these waves in order from lowest frequency (longest wavelength) to highest frequency (shortest wavelength):

__3__ blue visible light

__1__ AM radio waves

__4__ x rays

__2__ infrared light.

28. Label the elementary particles of an atom:

electron
neutron
nucleus
proton
quark

Matching (1 Points Each Question)

1. ____ first law of motion
2. ____ second law of motion
3. ____ third law of motion
4. ____ force equation
5. ____ definition of impulse
6. ____ definition of momentum

a. $a = f/m$
b. $f = m \times a$
c. $f_{ab} = -f_{ba}$
d. $I = f \times t$
e. If $f = 0$ then $a = 0$
f. $p = m \times v$

7. ____ Archimedes' principle of buoyancy
8. ____ Boyle's law
9. ____ Ideal gas law
10. ____ Bernoulli's principle

a. Pressure times volume of any gas divided by the temperature is a constant.
b. The lifting force acting on a solid object immersed in water is equal to the weight of the water shoved aside by the object.
c. The velocity of a fluid and its pressure are inversely related.
d. The volume of a gas is inversely proportional to the pressure.

11. ____ brings light to a focus.
12. ____ controls the amount of light that enters the eye.
13. ____ is the opening through which light enters the eye.
14. ____ adjusts light to the best focus.
15. ____ is a surface of light sensitive nerves.
16. ____ carries information from the eye to brain.
17. ____ is sensitive to light but cannot see color.
18. ____ is sensitive to light and can distinguish color.

a. Cones
b. Cornea
c. Iris
d. Lens
e. Optic nerve
f. Pupil
g. Retina
h. Rods

19. ____ Niels Bohr
20. ____ Louis de Borglie
21. ____ Max Planck
22. ____ Werner Heisenberg

a. developed a model of the atom and electron orbits
b. explained black body radiation by using energy quanta
c. developed the uncertainty principle
d. proposed matter waves and calculated their wavelengths

Fill-in-the-Blank Questions (2 Points Each Answer)

23. To calculate speed, divide distance by time.

24. Momentum is the mass of an object times its velocity.

25. James Prescott Joule found how mechanical energy due to motion compares to heat energy.

26. The three factors that determine the heat contained in an object are type of substance, mass, and tempurcatur.

27. The three properties of a sound are frequency, intensity, and quality.

28. To reduce the heating effect of electricity in wires, the current is reduced but the Voltage is increased.

29. The Heisenberg uncertainty principle states that the precise position, mass, and Volocity cannot be determined exactly.

Multiple Choice Questions (2 Points Each Question)

30. Physics is the science that explores how energy acts on
    A. heat B. light (C.) matter D. sound

31. A feather and lump of lead will fall at the same speed in
    A. a high speed wind tunnel B. the atmosphere (C.) a vacuum D. water

32. Acceleration is found by dividing which of the following by the change in time?
    A. average velocity B. distance
    C. gravity (D.) change in speed

33. Inertia is a property of matter that resists changing its
    A. electric charge B. mass C. momentum (D.) velocity

34. Heat is a type of
    (A.) energy B. force C. matter D. temperature

35. The two most common substances used in thermometers are colored alcohol and
    A. cooking oil B. ethylene glycol (C.) mercury D. molten salt

36. The scientist who discovered that pure water has a fixed boiling and freezing temperature was
    A. Anders Celsius B. Antoine Lavoisier
    (C.) Daniel Fahrenheit D. John Dalton

37. Density is equal to mass divided by
    A. area B. pressure (C.) volume D. weight

38. The frequency of a sound is known as its
    A. amplitude B. color (C.) pitch D. velocity

39. The study of sound is known as
    (A.) acoustics B. astronomy C. mechanics D. thermodynamics

40. The loudness of sound is measured in
    A. candles (B.) decibels C. joules D. watts

41. The force that pushes electrons around a circuit is
A. resistance B. charge C. current D. voltage

42. The scientist who developed four equations that summarized electromagnetism was
A. Albert Einstein B. Isaac Newton
C. James Clerk Maxwell D. Michael Faraday

43. The first scientist to generate electromagnetic waves was
A. Arthur Compton B. Guglielmo Marconi
C. Michael Faraday D. Rudolf Hertz

44. The first person to state that uranium could undergo fission and produce a self-sustaining chain reaction was
A. Albert Einstein B. Enrico Fermi
C. Franklin D. Roosevelt D. Lise Meiter

45. The purpose of a moderator is to
A. absorb neutrons B. cool the reactor C. generate heat D. slow neutrons

Short Answer Questions (4 Points Each Question)

46. State the second law of motion:

47. State the third law of motion:

48. State Ohm's law: current is proporshonal to its voltage

True and False (2 Points Each)

49. T F Doubling mass of a moving object doubles its kinetic energy.

50. T F Doubling velocity of a moving object doubles its kinetic energy.

51. T F Scientists are unable to measure temperatures greater than 1,700°F.

52. T F Heat is the motion of atoms and molecules.

53. T F Thales of Melitus discovered that amber could be given a charge of static electricity.

54. T F An object with a positive charge has more protons than electrons.

Applied Learning Activities (1 Point Each Answer)

55. Label the four forces acting on an airplane in flight.

upward lift
drag
thrust
gravity

56. Write the numbers 1 to 4 in the blanks to rank these waves in order from lowest frequency (longest wavelength) to highest frequency (shortest wavelength):

__3__ blue visible light

__1__ AM radio waves

__4__ x rays

__2__ infrared light

Answer Keys

Chapter 1

1. F, 2. F
3. So the calendar will match the seasons. Or, so the calendar year will be the same length as the solar year.

4. a, 5. c, 6. e, 7. d, 8. a, 9. b

10. 969 years x 365 days per year = 353,685 days
11. 120 days, 72 days, 18 days, 6 days (divide 360 by 3, 5, 20, and 60)
12. divide the population by 1,461

Chapter 2

1. A, 2. F, 3. D, 4. A, 5. F,
6. C, 7. D, 8. B, 9. B, 10. B

11. 6:30 a.m. The second watch began at 4:00 a.m. Each bell is ½ hour. Five bells are 2½ hours: 4:00 a.m. + 2 hr. 30 min. = 6:30 a.m.
12. Answer varies depending on actual heart rate. For 72 beats per min.: 72 beats per min. x 60 min per hr. x 24 hr. per day = 103,680 beats per day
13. 8 hours. One way to solve the problem is to change to military time and subtract — 9:00 a.m. is 0900 and 5:00 p.m. is 1700: 1700 – 0900 = 0800 or 8 hours.
14. one hour later, 4:00 p.m. MST is 5:00 p.m. CST

Chapter 3

1. B, 2. A, 3. C, 4. A, 5. F
6. B, 7. 5,280, 8. pound, 9. B
10. A, 11. A, 12. B, 13. A

14. 60 inches, 5 feet. Multiplying 15 hands by 4 inches per hand gives 60 inches. Sixty inches is equal to five feet: 60 in ÷ 12 in. per ft. = 5 ft.
15. Answer varies. Multiply weight in pounds by the conversion factor of 16 ounces per pound.
16. 5.499 miles or about 5.5 miles. Divide 29,035 feet by the conversion factor of 5,280 feet per mile.

Chapter 4

1. B, 2. B, 3. F, 4. D, 5. T
6. B, 7. B, 8. B, 9. A, 10. D
11. D, 12. F, 13. C

Chapter 5

1. B, 2. T, 3. F, 4. B, 5. D
6. F, 7. A, 8. D, 9. A, 10. A
11. e, 12. a, 13. d , 14. c, 15. b
16. d, 17. e, 18. f, 19. b, 20. a
21. c

22. 140 tiles. The area of the room is 140 square feet, A = L x W = 14 ft. x 10 ft. = 140 sq. ft., and each tile covers one square foot, so 140 tiles are needed.

Chapter 6

1. B
2. squares, square
3. A, 4. D, 5. A, 6. e, 7. d, 8. c
9. a, 10. b, 11. b, 12. d, 13. a
14. c

Chapter 7

1. T, 2. F, 3. T, 4. T, 5. F, 6. T
7. T, 8. F, 9. F, 10. F, 11. F
12. F, 13. F

Chapter 8

1. B, 2. C, 3. A, 4. A, 5. A
6. C, 7. B, 8. A, 9. B, 10. F
11. T, 12. B, 13. C
14. 233 = 89 + 144
15. F, 16. b, 17. c, 18. a, 19. d
20. e

Chapter 9

1. B, 2. B, 3. T, 4. T, 5. F, 6. T
7. D, 8. C, 9. B, 10. A, 11. A
12. F, 13. B, 14. D, 15. a, 16. g

17. b, 18. e, 19. c, 20. f, 21. d

Chapter 10

1. T, 2. C, 3. A, 4. B, 5. T, 6. F

7. C, 8. B, 9. C, 10. b, 11. d

12. c, 13. a, 14. b, 15. d, 16. a

17. c

Chapter 11

1. A, 2. A, 3. C, 4. B, 5. B

6. C, 7. B, 8. F, 9. e, 10. b

11. a, 12. d, 13. c

14. 17,576,000 — Any one of 26 letters (A through Z) can be chosen to fill the first three positions. Any one of 10 digits (zero through nine) can be chosen to fill the second group of three positions: 26 x 26 x 26 x 10 x 10 x 10 = 17,576,000.

Chapter 12

1. B, 2. F, 3. T, 4. F, 5. T

6. three, ten, 0.477

7. D, 8. B, 9. T, 10. B

11. central processing unit

12. A, 13. h, 14. c, 15. d, 16. f

17. g, 18. a, 19. e, 20. b

Chapter 13

1. 9, 2. F, 3. A, 4. B, 5. C, 6. T

7. B, 8. D, 9. F, 10. F, 11. D

12. B, 13. B, 14. A, 15. C, 16. T

17. A

18. About 3.8 seconds: 26,747 characters / 7,000 bytes (characters) per second = 3.821 seconds

Chapter 14

Puzzle 1: Multiplying by Seven. You discover that 7 x 142,857 = 999,999. The reason that the answer suddenly goes to all nines is surprising, but only because the source of the number 142,857 was not given. The number 100,000 divided by 7 is a repeating decimal: 1,000,000 ÷ 7 = 142,857.142857142857 . . . with the pattern 142857 repeating. Now 1,000,000 ÷ 7 x 7 = 1,000,000, so 7 x 142,857.142857142857 . . . would also be 1,000,000. However, if only the first group is used, then 7 x 142,857 = 999,999. The final one needed to roll the number up to one million is missing because the repeating part of the decimal fraction was not used.

Puzzle 2: Multiplying by 99. The left-most digit (the one in the 100s place) in the answer goes from 1 to 8 while the right-most digit (the one in the 1s place) goes from 8 to 1. This result is easy to see once you think that the number 99 is 100 – 1. Multiplying by two gives 200 – 2 = 198; by three gives 300 – 3 = 297; by four gives 400 – 4 = 396 and so on.

Puzzle 3: On the Road to St. Ives. Only one person is on the road to St. Ives. Instead of a large number, this is a trick question. The last line of the rhyme asks how many were going to St. Ives. The wives and the things they carried were going away from St. Ives. The narrator (the "I" person) is the one going to St. Ives.

Puzzle 4: Send More Money.

```
  S E N D
+ M O R E
M O N E Y

  9 5 6 7
+ 1 0 8 5
1 0 6 5 2
```

M in MONEY must be 1 because even with a carry, the sum of S and M is less than 20. O must be zero because M is 1 and S must be 9 or 8 with a carry of 1. Either value for S forces O to be zero. Now that we know O is zero, S must be 9 because 8 is too small even with a carry of 1 to be ten.

The total of N and R must be greater than 10 to give a carry of 1; otherwise E plus zero and no carry would equal R, and two letters cannot have the same value. N is one more than E. Trying the remaining numbers in E + 0 = N and N + R = 10 + E shows that R is 8, N is 6 and E is 5. D = 7, E = 5, M = 1, N = 6, O = 0, R = 8, S = 9, Y = 2.

Puzzle 5: For More Study — The 3N + 1 Problem. The sequence is 18, 9, 28, 14, 7, 22, 11, 34, 17, 52, 26, 13, 40, 20, 10, 5, 16, 8, 4, 2, 1.

Puzzle 6: Samson's Riddle. Judges 14:14 tells how Samson came to think of the puzzle. He found a lion's carcass with a honeycomb from a beehive inside. Out of the eater (lion), something to eat, out of the strong, something sweet (honey).

Puzzle 7: Sock Puzzle. One extra sock is enough. If the socks he is wearing match, he does not need the spare one. If his socks are not alike, then one will be the same color as the one he is carrying, giving him a matching pair.

Puzzle 8: River Crossing. First, take the goat across and leave him on the far bank. The carrots will be safe left alone with the wolf. Return for the carrots and leave them on the far bank but pick up the goat and return with him to the near bank. Leave the goat and paddle the wolf across. Leave the wolf with the carrots and return for the goat.

Puzzle 9: Dürer's Number Square. To get started, you should figure out the sum for each row, column, and diagonal. The numbers 1 through 9 sum to 45, and each row (and column and diagonal) must sum to 15 (45/3 = 15). One possible arrangement is:

8	1	6
3	5	7
4	9	2

The other seven solutions are merely rotations and mirror images of this solution.

Puzzle 10: Grass to Milk. BESSIE is the cow's name.

Multiply prime numbers: 7 x 17 x 23 = 2,737.

Find area: A = L x W = 201 ft. x 201 ft. = 40,401 sq. ft.

Seconds in a day: 60 sec/min. x 60 min./hr. x 24 hr./day = 86,400 sec.

Speed of light 186,000 mi./sec.

Adding the numbers (ignore the units): 2,737 + 40,401 + 86,400 + 186,000 = 315538.

Exploring the World of Physics — Worksheet Answer Keys

Chapter 1

1. C. matter
2. F. They seldom did experiments.
3. T
4. F. length and weight
5. C. a vacuum
6. time
7. D. rolled them down a ramp
8. D. change in speed
9. A. 32 ft./sec^2
10. 53 miles per day. Average speed is the distance divided by the time: speed = distance/time = 3,710 miles/70 days = 53 mi./da.
11. 7.5 mi./hr. × sec. Acceleration = (change in speed)/(change in time) = (60 mi./hr.) (8 sec.) = 7.5 mi./hr. × sec.
12. 31.8 ft./sec. $v_f = a \times t$ = (5.3 ft./sec^2) (6 sec.) = 31.8 ft./sec.

Chapter 2

1. F. velocity includes direction as well as speed
2. T

3. F. force must act on a moving object to slow it to a stop
4. A. friction
5. B. Galileo
6. D. velocity
7. F. every object has inertia
8. T
9. The acceleration of an object is directly proportional to the force acting on it and inversely proportional to its mass.
10. To every action there is an equal and opposite reaction.
11. velocity

12. T, 13. e, 14. a, 15. c, 16. b
17. d, 18. f

Chapter 3

1. T
2. B — elliptical
3. A — faster
4. The straight line joining a planet with the sun sweeps out equal areas in equal intervals of time.
5. T
6. B — Kepler
7. F — his father had died and his mother was poor
8. T
9. A — 3,600 times weaker ($60^2 = 3,600$)
10. F. to all objects
11. B
12. product, square
13. F. Scientists believe they have found about 100 planets around distant stars.

Chapter 4

1. force
2. A. Archimedes
3. A. effort
4. B. a lever
5. F. at either end or anywhere in between
6. B. reduced
7. M.A. = 9. Mechanical advantage = ramp length/ ramp height = run rise = 9 miles/1 mile = 9
8. C. a wheel and axle
9. T
10. F — it is less
11. B. 18 wheeler truck
12. B. 100

Chapter 5

1. F — only since the 1800s
2. B — energy
3. T
4. D — work
5. energy
6. D — work
7. heat
8. F — The desk must move for work to be done.
9. B — power
10. C — horsepower and watt
11. A — kinetic
12. T
13. F — it becomes four times as great
14. B — doubling its velocity
15. B — potential energy
16. A - heat

Chapter 6

1. A — energy
2. temperature
3. B — water
4. B — expand
5. C — mercury
6. F — They can measure higher temperatures with electrical conductivity and color of light emitted by glowing substance.
7. C — Daniel Fahrenheit
8. A — higher
9. T
10. A — kinetic
11. A — conduction

12. A — copper
13. A — poorly
14. B — convection
15. T
16. T
17. B — greatly different
18. T
19. 0.058 or about six percent
Efficiency = $(T_1 - T_2)/T_1$ = (291 K - 274 K)/(291 K) = 17/291 = 0.058 or less than six percent

Chapter 7

1. F — Because it will snap back to its original shape.
2. T
3. force
4. B — reduces
5. A — height
6. T
7. C — volume
8. B — decrease
9. B — square root
10. b, 11. d, 12. a, 13. c

Chapter 8

1. T
2. D — wavelength
3. A — frequency
4. B — velocity
5. wavelength
6. T
7. C — pitch
8. A — amplitude
9. A — high
10. F — Both travel at the same speed
11. tension
12. B — lowest
13. A — acoustics
14. A — better
15. B — twice
16. B — decibels
17. T
18. quality
19. A — same speed
20. A — toward

Chapter 9

1. b, 2. c, 3. f, 4. d, 5. g, 6. e
7. h, 8. a, 9. T
10. blue
11. A — reflection
12. B — virtual
13. B — mirror
14. A — convex
15. B — slower
16. A — refraction
17. F — color

Chapter 10

1. T
2. A — electron
3. T
4. charge
5. A — electrostatic
6. B — nonconductor
7. T
8. F — used in golf carts and hybrid cars
9. D — voltage
10. current = voltage/resistance, or in words: current is directly proportional to voltage and inversely proportional to resistance
11. C — resistance
12. voltage

Chapter 11

1. F — Understanding of magnetism was filled with misinformation
2. F — It is drawn toward the north magnetic pole of the earth.
3. F — Magnetic north is 1,100 miles from geographic north.

4. B — gravity
5. T, 6. T
7. C — iron
8. B — repel
9. T, 10. T
11. B — steel
12. A — lost
13. T
14. A — can be turned on and off
15. T
16. D — all of the above

Chapter 12

1. C — James Clerk Maxwell
2. C — the same
3. D — Rudolf Hertz
4. F — FM waves are too short to reflect
5. A — AM
6. C — modulation
7. 3–blue visible light; 1 — AM radio waves; 4 — x-rays; 2 — infrared light
8. D — the second scientific revolution
9. F — color of the light (frequency)
10. T
11. C — photoelectric effect
12. F — mass and velocity
13. T
14. F — Compton effect demonstrates particle nature of light, too.

Chapter 13

1. A — electron
2. B — neutron
3. B — a proton
4. A — fission
5. T
6. C — neutrons
7. D — Lise Meiter
8. T
9. D — slow neutrons
10. A — less
11. F — about 20 percent
12. F — Cold fusion is an unsolved problem.

Chapter 14

1. A — far less
2. F — its frequency
3. A — more
4. T
5. A — an electron
6. T
7. A — fundamental frequency
8. B — an electron
9. T
10. velocity
11. a, 12. d, 13. b, 14. c

Exploring the World of Mathematics — Quiz Answer Keys

Unit One Quiz, chapters 1–4

1. c. earth rotates on its axis once
2. e. seven days
3. d. moon revolves around the earth once
4. a. due to the tilt of the earth's axis, equal to three months
5. b. earth revolves around the sun once
6. 5,280
7. pound
8. A. authorized by Julius Caesar
9. D. Romans
10. C. 2400
11. D. trains
12. B. engineers used two different measures of force
13. C. drugs

14. B. human body,
15. D. the distance light travels in 1/299,792,458 of a second
16. D. 212 degrees
17. (divide 360 by 3, 5, 20, and 60), a. 120 days, b. 72 days, c. 18 days, d. 6 days
18. a. 60 inches (multiplying 15 hands by 4 inches per hand gives 60 inches)

    b. 5 feet (60 inches is equal to 5 feet: 60 in ÷ 12 in. per ft. = 5 ft.)
19. So the calendar will match the seasons; or so the calendar year will be the same length as the solar year.
20. 6:30 a.m. The second watch began at 4:00 a.m. Each bell is ½ hour. Five bells are 2½ hours: 4:00 a.m. + 2 hr. 30 min. = 6:30 a.m.
21. one hour later, 4:00 p.m. MST is 5:00 p.m. CST
22. 5.499 miles or about 5.5 miles. Divide 29,035 feet by the conversion factor of 5,280 feet per mile.
23. Answer varies depending on actual heart rate. For 72 beats per min: 72 beats per min. x 60 min. per hr. x 24 hr. per day = 103,680 beats per day
24. B 25. A 26. A 27. B 28. A

Unit Two Quiz, chapters 5–8

1. e. is not a polygon
2. a. a polygon with five sides
3. d. a quadrilateral with opposite sides parallel and equal in length
4. c. a polygon with three sides and one right angle
5 b. a rectangle with four equal sides
6. d. perimeter of a rectangle
7. e. perimeter of a square
8. f. volume
9. b. area of a rectangle
10. a. area of a circle
11. c. area of a square
12. e. ancient Greek who worked out a way to show large numbers that he called myriads
13. d. wrote Elements of Geometry
14. c. proved planets follow elliptical orbits
15. a. discovered that the sum of the 3 angles of any triangle is 180 degrees
16. b. used ratios to find the heights of buildings
17. b. all points are the same distance from the center
18. d. the orbit of Halley's comet is of this shape
19. a. a mirror of this shape will focus sunlight
20. c. the first part of the name means over or beyond
21. b. palindromes
22. c. prime numbers
23. a. Fibonacci numbers 8
24. d. square numbers
25. e. triangular numbers
26. squares, square
27. 233 = 89 + 144
28. D. a triangle with a right angle
29. D. eight times as great
30. D. an ellipse
31. C. number theory
32. F 33. F 34. F 35. F 36. F
37. 140 tiles. The area of the room is 140 square feet, A = L x W = 14 ft. x 10 ft. = 140 sq. ft., and each tile covers one square foot, so 140 tiles are needed.
38. 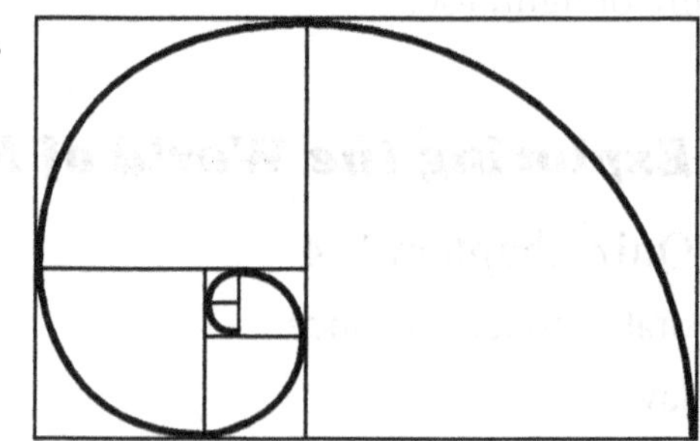

Unit Three Quiz, chapters 9–11

1. a. counting numbers
2. g. whole numbers
3. b. even numbers
4. e. odd numbers
5. c. integers
6. f. rational numbers
7. d. irrational numbers

8. b, 9. d, 10. c, 11. a, 12. b

13. d, 14. a, 15. c

16. e. proved that xn + yn = zn has no solution with whole numbers except for n = 2

17. b. he called his triangle an arithmetic triangle

18. a. discovered how to calculate the coefficients of a binomial raised to a power

19. d. solved the Konigsberg bridge problem

20. c. his last theorem was solved in 1995

21. D. division

22. B. 4

23. B. 25

24. A. denominator

25. B. an irrational number

26. D. all of the above

27. C. solving

28. A. a coefficient

29. B. geometry

30. C. four-color map problem

31. B. factorial

32. C. 120

33. 17,576,000 — Any one of 26 letters (A through Z) can be chosen to fill the first three positions. Any one of 10 digits (zero through nine) can be chosen to fill the second group of three positions: 26 x 26 x 26 x 10 x 10 x 10 = 17,576,000.

34. a. Utah, b. Colorado, c. Arizona, d. New Mexico

Unit Four Quiz, chapters 12–14

1. h. wrote the first computer program

2. c. built first general purpose calculating machine

3. d. built the difference engine

4. f. invented tabulating machines used in the 1890 census

5. g. spent six years calculating the orbit of Mars,

6. a. built a calculator called the Step Reckoner

7. e. invented logarithms

8. b. built a calculator to help his father, a tax collector

9. three, ten, 0.477

10. 9

11. B. 1

12. D. 5.28 x 103

13. B. punched cards

14. A. random access memory

15. A. bit

16. C. eight

17. D. the three colors of red, green, and blue

18. D. transmitting only those pixels that are different from the previous one

19. B. months

20. A. electric signals can go no faster than the speed of light

21. central processing unit

22. About 3.8 seconds: 26,747 characters / 7,000 bytes (characters) per second = 3.821 seconds

23. Samson's Riddle: Judges 14:14 tells how Samson came to think of the puzzle. He found a lion's carcass with a honeycomb from a beehive inside. Out of the eater (lion), something to eat, out of the strong, something sweet (honey).

24. On the road to St. Ives: Only one person is on the road to St. Ives. Instead of a large number, this is a trick question. The last line of the rhyme asks how many were going to St. Ives. The wives and the things they carried were going away from St. Ives. The narrator (the "I" person) is the one going to St. Ives.

25. River Crossing: First, take the goat across and leave him on the far bank. The carrots will be safe left alone with the wolf. Return for the carrots and leave them on the far bank but pick up the goat and return with him to the near bank. Leave the goat and paddle the wolf across. Leave the wolf with the carrots and return for the goat.

Exploring the World of Mathematics Test Answer Key

1. c. earth rotates on its axis once
2. e. seven days
3. d. moon revolves around the earth once
4. a. due to the tilt of the earth's axis, equal to three months
5. b. earth revolves around the sun once
6. b. all points are the same distance from the center
7. d. the orbit of Halley's comet is of this shape
8. a. a mirror of this shape will focus sunlight
9. c. the first part of the name means over or beyond
10. a. counting numbers
11. g. whole numbers
12. b. even numbers
13. e. odd numbers
14. c. integers
15. f. rational numbers
16. d. irrational numbers
17. h. wrote the first computer program
18. c. built first general purpose calculating machine
19. d. built the difference engine
20. f. invented tabulating machines used in the 1890 census
21. g. spent six years calculating the orbit of Mars
22. a. built a calculator called the Step Reckoner
23. e. invented logarithms
24. b. built a calculator to help his father, a tax collector
25. 5,280
26. pound
27. squares, square
28. 233 = 89 + 144
29. three, ten, 0.477
30. 9
31. B. engineers used two different measures of force
32. C. drugs
33. D. the distance light travels in 1/299,792,458 of a second
34. D. 212 degrees
35. D. a triangle with a right angle
36. D. eight times as great
37. D. an ellipse
38. C. number theory
39. B. 25
40. A. denominator
41. A. a coefficient
42. B. geometry
43. D. the three colors of red, green, and blue
44. D. transmitting only those pixels that are different from the previous one
45. B. months
46. A. electric signals can go no faster than the speed of light
47. a. 60 inches (multiplying 15 hands by 4 inches per hand gives 60 inches.)
    b. 5 feet (60 inches is equal to 5 feet: 60 in. ÷ 12 in. per ft. = 5 ft.)
48. So the calendar will match the seasons; or so the calendar year will be the same length as the solar year.
49. one hour later, 4:00 p.m. MST is 5:00 p.m. CST
50. 5.499 miles or about 5.5 miles. Divide 29,035 feet by the conversion factor of 5,280 feet per mile.
51. 140 tiles. The area of the room is 140 square feet, A = L x W = 14 ft. x 10 ft. = 140 sq. ft.,

and each tile covers one square foot, so 140 tiles are needed.

52. Sock Puzzle: One extra sock is enough. If the socks he is wearing match, he does not need the spare one. If his socks are not alike, then one will be the same color as the one he is carrying, giving him a matching pair.

53. Durer's Number Square: To get started, you should figure out the sum for each row, column, and diagonal. The numbers 1 through 9 sum to 45, and each row (and column and diagonal) must sum to 15 (45/3 = 15). One possible arrangement is:

8	1	6
3	5	7
4	9	2

The other seven solutions are merely rotations and mirror images of this solution.

Exploring the World of Physics — Quiz Answer Keys

Unit One Quiz, chapters 1–4

1. e. If f = 0 then a = 0
2. a. a = f/m
3. c. fab = -fba
4. b. f = m x a
5. d. I = f x t
6. f. p = m x v
7. time
8. 53 miles per day. Average speed is the distance divided by the time: speed = distance/time = 3,710 miles/70 days = 53 mi/da.
9. 31.8 ft./sec. v1 = a x t = (5.3 ft./sec2) (6 sec.) = 31.8 ft/sec
10. Velocity
11. product, square
12. M.A. = 9
13. C. matter
14. C. a vacuum
15. D. rolled them down a ramp
16. D. change in speed by the change in time.
17. D. velocity
18. B. elliptical
19. A. Archimedes
20. B. a lever
21. 7.5 mi./hr. x sec. Acceleration = (change in speed)/(change in time) = (60 mi./hr.) (8 sec.) = 7.5 mi./hr. x sec.
22. The acceleration of an object is directly proportional to the force acting on it and inversely proportional to its mass.
23. To every action there is an equal and opposite reaction.
24. A. forward thrust of propeller
    B. drag
    C. upward lift of wings
    E. downward pull of gravity
25. A. Fulcrum
    B. Load
    C. Effort Points
26. A. Fulcrum
    B. Load
    C. Effort Points

Unit Two Quiz, chapters 5–7

1. b. The lifting force acting on a solid object immersed in water is equal to the weight of the water shoved aside by the object.
2. d. The volume of a gas is inversely proportional to the pressure.
3. a. Pressure times volume of any gas divided by the temperature is a constant.
4. c. The velocity of a fluid and its pressure are inversely related.
5. energy
6. heat

7. temperature
8. 0.058 or about six percent.

   Efficiency = (T1 – T2)/T1 = (291 K – 274 K)/(291 K) = 17/291 = 0.058 or less than six percent
9. D. work
10. D. work
11. A. energy
12. C. mercury
13. C. Daniel Fahrenheit
14. C. volume
15. C. horsepower and watt
16. A. heat
17. A. conduction
18. B. convection
19. B. square root
20. F — only since the 1800s
21. T
22. F — The desk must move for work to be done.
23. T
24. F — It becomes four times as great.
25. F — They can measure higher temperatures with electrical conductivity and color of light emitted by glowing substance.

26. T, 27. T, 28. T, 29. T

30. F — Because it will snap back to its original shape.

31. T, 32. T

33. You can measure heat capacity by heating identical masses of different substances and then placing them on a block of ice. Observe the different distances they melt into the ice.

Unit Three Quiz, chapters 8–10

1. b. Cornea
2. c. Iris
3. f. Pupil
4. d. Lens
5. g. Retina
6. e. Optic nerve
7. h. Rods
8. a. Cones
9. tension
10. quality
11. blue
12. charge
13. voltage
14. B. lowest
15. A. better
16. A. same speed
17. A. toward
18. A. reflection
19. B. virtual
20. B. mirror
21. A. convex
22. B. slower
23. refraction
24. A. electron
25. A. electrostatic
26. B. nonconductor
27. D. wavelength
28. C. pitch
29. A. acoustics
30. B. decibels
31. D. voltage
32. C. resistance
33. current = voltage/resistance, or in words: current is directly proportional to voltage and inversely proportional to resistance.

34. T, 35. T

36. F — color

37. T, 38. T, 39. T

40. F — used in golf carts and hybrid cars
41. rustling of leaves
42. whisper
43. average home sounds (such as the humming of a refrigerator)

44. automobile
45. ordinary conversation
46. heavy street traffic
47. jack hammer
48. thunder

Unit Four Quiz, chapters 11–14

1. a. Developed a model of the atom and electron orbits
2. d. Proposed matter waves and calculated their wavelengths
3. b. Explained black body radiation by using energy quanta
4. c. Developed the uncertainty principle
5. velocity
6. A. can be turned on and off
7. D. all of the above
8. C. James Clerk Maxwell
9. D. Rudolf Hertz
10. C. modulation
11. D. the second scientific revolution
12. C. photoelectric effect
13. A. an electron
14. C. neutrons
15. D. Lise Meiter
16. D. slow neutrons
17. B. an electron
18. B. steel
19. A. lost
20. A. AM
21. A. fission
22. A. less
23. A. far less
24. A. more
25. A. an electron
26. A. fundamental frequency
27. 3 — blue visible light; 1 — AM radio waves; 4 — x-rays; 2 — infrared light
28. A. nucleus
    B. electron
    C. proton
    D. neutron
    E. quark

Exploring the World of Physics Test Answer Key

1. e. If f = 0 then a = 0
2. a. a = f/m
3. c. fab = -fba
4. b. f = m x a
5. d. I = f x t
6. f. p = m x v
7. b. The lifting force acting on a solid object immersed in water is equal to the weight of the water shoved aside by the object.
8. d. The volume of a gas is inversely proportional to the pressure.
9. a. Pressure times volume of any gas divided by the temperature is a constant.
10. c. The velocity of a fluid and its pressure are inversely related.
11. b. Cornea
12. c. Iris
13. f. Pupil
14. d. Lens
15. g. Retina
16. e. Optic nerve
17. h. Rods
18. a. Cones
19. a. developed a model of the atom and electron orbits
20. d. proposed matter waves and calculated their wavelengths
21. b. explained black body radiation by using energy quanta

22. c. developed the uncertainty principle

23. time

24. Velocity

25. heat

26. temperature

27. quality

28. voltage

29. velocity

30. C. matter

31. C. a vacuum

32. D. change in speed

33. D. velocity

34. A. energy

35. C. mercury

36. C. Daniel Fahrenheit

37. C. volume

38. C. pitch

39. A. acoustics

40. B. decibels

41. D. voltage

42. C. James Clerk Maxwell

43. D. Rudolf Hertz

44. D. Lise Meiter

45. D. slow neutrons

46. The acceleration of an object is directly proportional to the force acting on it and inversely proportional to its mass.

47. To every action there is an equal and opposite reaction.

48. current = voltage/resistance, or in words: current is directly proportional to voltage and inversely proportional to resistance.

49. T

50. F — It becomes four times as great

51. F — They can measure higher temperatures with electrical conductivity and color of light emitted by glowing substance.

52. T , 53. T , 54. T

55. A. forward thrust of propeller

B. drag

C. upward lift of wings

E. downward pull of gravity

56. 3 — blue visible light; 1 — AM radio waves; 4 — x-rays; 2 — infrared light

Master Books®
Curriculum

Preschool

Biblical Beginnings

Ages 3, 4, & 5 years old

Package Includes: *A is for Adam; A is for Adam DVD; D is for Dinosaur; D is for Dinosaur DVD; N is for Noah; Noah's Ark Preschool; God Made the World & Me; Dinosaurs Stars of the Show; Big Thoughts for Little Thinkers: The Gospel; The Mission; The Scripture; The Trinity; Creation Story for Children; When Dragons Hearts Were Good; Dinosaur by Design; Biblical Beginnings Preschool , Parent Lesson Planner*

14 Book, 2 DVD Package 978-0-89051-887-8 $187.84
PLP Only (168 Pages) 978-0-89051-885-4 $14.99

Elementary

Elementary Bible & English Grammar

1 year: 3rd–6th grade

Package Includes: *Illustrated Family Bible Stories and Parent Lesson Planner*

2 Book Package 978-0-89051-852-6 $39.99
PLP Only (130 Pages) 978-0-89051-851-9 $14.99

Elementary Geography & Cultures

1 year: 3rd–6th grade

Package Includes: *Children's Atlas of God's World, Passport to the World, & Parent Lesson Planner*

3 Book 978-0-89051-814-4 $49.99
PLP Only (130 Pages) 978-0-89051-808-3 $14.99

Elementary World History

1 year: 5th–8th grade

Package Includes: *The Big Book of History; Noah's Ark: Thinking Outside the Box (book and DVD); Parent Lesson Planner*

3 Book, 1 DVD Package 978-0-89051-815-1 $64.99
PLP Only (260 Pages) 978-0-89051-809-0 $15.99

Elementary Zoology

1 year: 3th–6th grade

5 Book Package Includes: *World of Animals; Dinosaur Activity Book; The Complete Aquarium Adventure; The Complete Zoo Adventure; Parent Lesson Planner*

5 Book Package 978-0-89051-747-5 $85.99
PLP Only (178 Pages) 978-0-89051-724-6 $14.99

Elementary

Science Starters: Elementary Chemistry & Physics

1 year: 3rd–6th grade

7 Book Package Includes: *Matter- Student, Student Journal, and Teacher; Energy- Student, Teacher, and Student Journal; Parent Lesson Planner*

7 Book Package	978-0-89051-749-9	$54.99
PLP Only (100 Pages)	978-0-89051-726-0	$8.99

Science Starters: Elementary General Science & Astronomy

1 year: 3rd–6th grade

7 Book Package Includes: *Water & Weather- Student Journal and Teacher; The Universe- Student, Teacher, & Student Journal; Parent Lesson Planner*

7 Book Package	978-0-89051-816-8	$54.99
PLP Only (98 Pages)	978-0-89051-810-6	$8.99

Science Starters: Elementary Physical & Earth Science

1 year: 3rd–6th grade

6 Book Package Includes: *Forces & Motion- Student, Student Journal, and Teacher; The Earth- Student, Teacher & Student Journal; Parent Lesson Planner*

6 Book Package	978-0-89051-748-2	$51.99
PLP Only (94 Pages)	978-0-89051-725-3	$8.99

Middle School

Applied Engineering: Studies of God's Design In Nature

1 year Engineering (1 credit): 7th–9th grade

Package Includes: *Made in Heaven, Champions of Invention, Discovery of Design, Parent Lesson Planner*

4 Book Package	978-0-89051-812-0	$53.99
PLP Only (260 Pages)	978-0-89051-806-9	$17.99

Christian Heritage

1 year Christian Heritage (1 credit): 9th–12th grade

Package Includes: *For You They Signed; Parent Lesson Planner*

2 Book Package	978-0-89051-769-7	$50.99
PLP Only (240 Pages)	978-0-89051-746-8	$15.99

Christian History: Biographies of Faith

1 year History (1 credit): 7th–9th grade

6 Book Package Includes: *Life of John Newton, Life of Washington, Life of Andrew Jackson, Life of John Knox, Life of Luther, Parent Lesson Planner*

6 Book Package	978-0-89051-847-2	$101.99
PLP Only (176 Pages)	978-0-89051-850-2	$16.99

High School

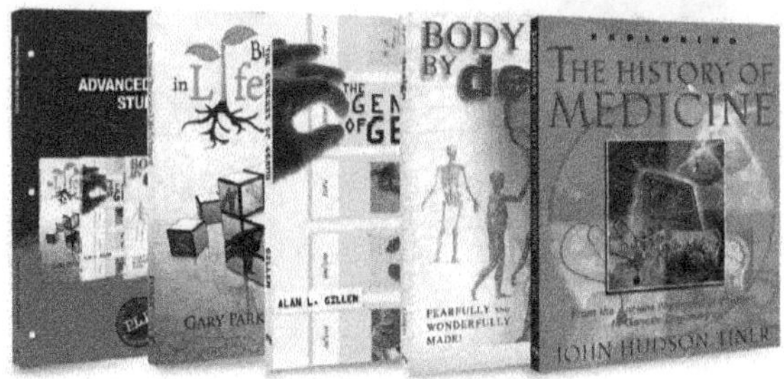

Advanced Pre-Med Studies

1 year Biology (1 credit): 10th–12th grade

Package Includes: *Building Blocks in Life Science; The Genesis of Germs; Body by Design; Exploring the History of Medicine; Parent Lesson Planner*

5 Book Package 978-0-89051-767-3 $79.99
PLP Only (238 pages) 978-0-89051-744-4 $17.99

Apologetics in Action

1 year Apologetics (1 credit): 10th–12th grade

Package Includes: *How Do I Know the Bible is True volumes 1 & 2; Demolishing Supposed Bible Contradictions Volumes 1 & 2; Parent Lesson Planner*

5 Book Package 978-0-89051-848-9 $70.99
PLP Only (176 Pages) 978-0-89051-839-7 $14.99

Basic Pre-Med

1 year Biology (½ credit): 8th–9th grade

Package Includes: *The Genesis of Germs; The Building Blocks in Life Science; Parent Lesson Planner*

3 Book Package 978-0-89051-759-8 $45.99
PLP Only (120 Pages) 978-0-89051-736-9 $12.99

Biblical Archaeology

1 year Bible/Archaeology (1 credit): 10th–12th grade

Package Includes: *Unwrapping the Pharaohs; Unveiling the Kings of Israel; The Archaeology Book; Parent Lesson Planner*

4 Book Package 978-0-89051-768-0 $99.99
PLP Only (244 Pages) 978-0-89051-745-1 $17.99

Cultural Issues: Creation/Evolution and the Bible

1 year Apologetics (½ credit): 10th–12th grade

Package Includes: *New Answers Book 1; New Answers Book 2; Parent Lesson Planner*

3 Book Package 978-0-89051-846-5 $44.99
PLP Only (208 Pages) 978-0-89051-849-6 $14.99

Intro to Biblical Greek

1 year Foreign Language (½ credit): 10th–12th grade

Package Includes: *It's Not Greek to Me DVD & Parent Lesson Planner*

1 Book, 1 DVD Package 978-0-89051-818-2 $33.99
PLP Only (132 Pages) 978-0-89051-817-5 $13.99

Intro to Economics: Money, History, & Fiscal Faith

1/2 year Economics (½ credit): 10th–12th grade

Package Includes: *Bankruptcy of Our Nation, Money Wise DVD, Parent Lesson Planner*

2 Book, 4 DVD Package 978-0-89051-811-3 $57.99
PLP Only (230 Pages) 978-0-89051-805-2 $13.99

Life Science: Origins & Scientific Theory

1 year Palentology (1 credit): 7th–9th grade

Package Includes: *Evolution: the Grand Experiment, Teacher Guide, DVD; Living Fossils, Teacher Guide, DVD; Parent Lesson Planner*

5 Book, 2 DVD Package 978-0-89051-761-1 $144.99
PLP Only (60 Pages) 978-0-89051-738-3 $5.99

Natural Science: The Story of Origins

1 year Natural Science (½ credit): 10th–12th grade

Package Includes: *Evolution: The Grand Experiment; Evolution: The Grand Experiment Teacher's Guide, Evolution: The Grand Experiment DVD; Parent Lesson Planner*

3 Book, 1 DVD Package 978-0-89051-762-8 $71.99
PLP Only (35 Pages) 978-0-89051-739-0 $4.99

Paleontology: Living Fossils

1 year Paleontology (½ credit): 10th–12th grade

Package Includes: *Living Fossils, Living Fossils Teacher Guide, Living Fossils DVD; Parent Lesson Planner*

3 Book, 1 DVD Package 978-0-89051-763-5 $66.99
PLP Only (31 Pages) 978-0-89051-740-6 $4.99

Survey of Astronomy

1 year Astronomy (1 credit): 10th–12th grade

Package Includes: *The Stargazers Guide to the Night Sky; Our Created Moon; Taking Back Astronomy; Our Created Moon DVD; Created Cosmos DVD; Parent Lesson Planner*

4 Book, 2 DVD Package 978-0-89051-766-6 $113.99
PLP Only (250 Pages) 978-0-89051-743-7 $17.99

Survey of Science History & Concepts

1 year Science (1 credit): 10th–12th grade

Package Includes: *The World of Mathematics; The World of Physics; The World of Biology; The World of Chemistry; Parent Lesson Planner*

5 Book Package 978-0-89051-764-2 $72.99
PLP Only (216 Pages) 978-0-89051-741-3 $16.99

Elementary

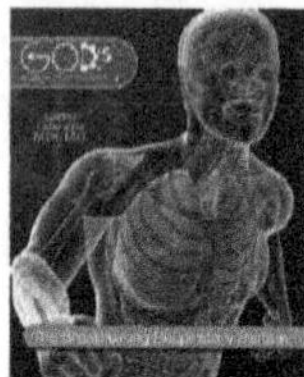

Breathtaking Respiratory System, The

Dr. Lainna Callentine

An elementary-level exploration of the human body's respiratory system, focused on structures, function, diseases, and God's design. Created by pediatrician and homeschool mom, Dr. Lainna Callentine.

C 978-0-89051-862-5 $15.99 U.S.

Electrifying Nervous System, The

Dr. Lainna Callentine

Learn interesting and important facts about why you sleep, what foods can superpower your brain functions, and much more in a wonderful exploration of the brain and how it controls your body!

C 978-0-89051-833-5 $15.99 U.S.

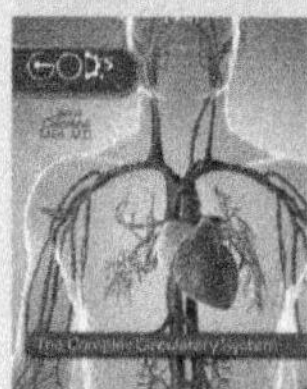

Complex Circulatory System, The

Dr. Lainna Callentine

Focuses on the heart, blood, and blood vessels that make up the body's circulatory system. Beyond the basics of how and why the body works, students learn God's amazing and deliberate design.

C 978-0-89051-908-0 $15.99 U.S.

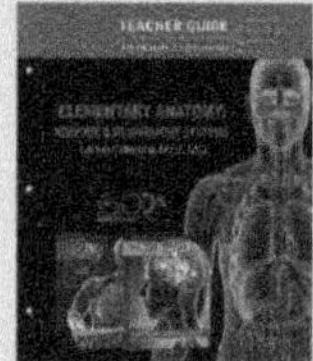

Elementary Anatomy Teacher Guide

Dr. Lainna Callentine

Instructional guide for 36-week elementary anatomy course based on the nervous and the respiratory systems, including weekly calendar, worksheets, activities, and tests focusing on the major concepts.

P 978-0-89051-842-7 $17.99 U.S.

The Fight for Freedom

Rick and Marilyn Boyer

This third-grade history course introduces readers to 18 heroes of early American history, and some villains as well. Students will learn of God's providential acts with this daily, 34-week curriculum.

STUDENT: P 978-0-89051-909-7 $29.99 U.S.
TEACHER: P 978-0-89051-912-7 $14.99 U.S.

America's Struggle to Become a Nation

Rick and Marilyn Boyer

This fourth-grade course provides a thorough understanding of the foundations of American government. Students learn of the War of Independence through the Constitution in this 34-week study.

STUDENT: P 978-0-89051-910-3 $29.99 U.S.
TEACHER: P 978-0-89051-911-0 $14.99 U.S.

Middle School

Studies in World History 1

James P. Stobaugh

Middle school history that covers the Fertile Crescent, Egypt, India, China, Japan, Greece, Christian history, and more. Begins with creation and moves forward with a solid biblically based worldview

STUDENT: P 978-0-89051-784-0 $29.99 U.S.
TEACHER: P 978-0-89051-791-8 $24.99 U.S. 

Skills for Language Arts

James P. Stobaugh

From the basics of grammar to a voyage through classic literature, this 34-week, junior high course lays a foundation for students who are serious about communicating their message. Five instructive lessons weekly.

STUDENT: P 978-0-89051-859-5 $34.99 U.S.
TEACHER: P 978-0-89051-860-1 $24.99 U.S.

Studies in World History 2

James P. Stobaugh

Middle school history that covers the clash of cultures, Europe and the Renaissance, Reformation, revolutions, and more. A comprehensive examination of history, including geography, economics, and government systems.

STUDENT: P 978-0-89051-785-7 $29.99 U.S.
TEACHER: P 978-0-89051-792-5 $24.99 U.S. 

Skills for Literary Analysis

James P. Stobaugh

Equips middle school students to analyze classic literary genres, discern authors' worldviews, and apply biblical standards.

STUDENT: P 978-0-89051-712-3 $34.99 U.S.
TEACHER: P 978-0-89051-713-0 $15.99 U.S.

Studies in World History 3

James P. Stobaugh

An entire year of high school American history curriculum in an easy-to-teach and comprehensive volume by respected Christian educator Dr. James Stobaugh.

STUDENT: P 978-0-89051-786-4 $29.99 U.S.
TEACHER: P 978-0-89051-793-2 $24.99 U.S.

Skills for Rhetoric

James P. Stobaugh

Help middle school students develop the skills necessary to communicate more powerfully through writing and to articulate their thought clearly.

STUDENT: P 978-0-89051-710-9 $34.99 U.S.
TEACHER: P 978-0-89051-711-6 $15.99 U.S.

Principles of Mathematics Book 1

Katherine A. Loop

Focus is on multiplication, division, fractions, decimals, ratios, percentages, shapes, basic geometry, and more, teaching clearly how math is a real-life tool pointing us to God and helping us explore His creation.

STUDENT: P 978-0-89051-875-5 $34.99 U.S.
Student Workbook: P 978-0-89051-876-2 $24.99 U.S.

Principles of Mathematics Book 2

Katherine A. Loop

Focus is on the essential principles of algebra, coordinate graphing, probability, statistics, functions, and other important areas of mathematics. Here at last is a curriculum with a biblical worldview.

STUDENT: P 978-0-89051-906-6 $34.99 U.S.
STUDENT WORKBOOK: P 978-0-89051-907-3 $29.99 U.S.

Middle School

Introduction to Anatomy & Physiology: Musculoskeletal System
Dr. Tommy Mitchell

An exploration of the awe-inspiring creation that is the human body. Explore the structure, function, and regulation of the body in detail, our bodies God created that are delicate, powerful, and complex.

C 978-0-89051-865-6 $17.99 U.S.

High School

American History
James P. Stobaugh

An entire year of high school American history curriculum in an easy-to-teach and comprehensive volume by respected Christian educator Dr. James Stobaugh.

STUDENT: P 978-0-89051-644-7 $29.99 U.S.
TEACHER: P 978-0-89051-643-0 $19.99 U.S.

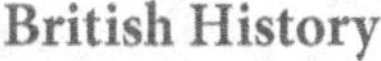

American Literature
James P. Stobaugh

A well-crafted high school presentation of whole-work selections from the major genres of American literature (prose, poetry, and drama), with background material on the writers and their worldviews.

STUDENT: P 978-0-89051-671-3 $39.99 U.S.
TEACHER: P 978-0-89051-672-0 $19.99 U.S.

British History
James P. Stobaugh

Examine historical theories, concepts, and global influences of this tiny country during an entire year of high school British history curriculum in an easy-to-teach and comprehensive volume.

STUDENT: P 978-0-89051-646-1 $24.99 U.S.
TEACHER: P 978-0-89051-645-4 $19.99 U.S.

British Literature
James P. Stobaugh

A well-crafted presentation of whole-work selections from the major genres of British literature (prose, poetry, and drama), with background material on the writers and their worldviews. High School.

STUDENT: P 978-0-89051-673-7 $34.99 U.S.
TEACHER: P 978-0-89051-674-4 $19.99 U.S.

World History
James P. Stobaugh

This study will help high school students develop a Christian worldview while forming his or her own understanding of world history trends, philosophies, and events.

STUDENT: P 978-0-89051-648-5 $24.99 U.S.
TEACHER: P 978-0-89051-647-8 $19.99 U.S.

World Literature
James P. Stobaugh

A well-crafted presentation of whole-work selections from the major genres of world literature (prose, poetry and drama), with background material on the writers and their worldviews. High school

STUDENT: P 978-0-89051-675-1 $34.99 U.S.
TEACHER: P 978-0-89051-676-8 $19.99 U.S.

ACT & College Preparation Course for the Christian Student
James P. Stobaugh

Your ACT score is key in determining college scholarships and admissions. Prepare to excel with these 50 devotion-based lessons.

P 978-0-89051-639-3 $29.99 U.S.

SAT
James P. Stobaugh

Written by a certified SAT grader, get insight designed just for Christian students to be well prepared for the test and score higher. Also focuses on spiritual disciplines of Bible reading and prayer.

P 978-0-89051-624-9 $29.99 U.S.

Building Blocks in Earth Science
Gary Parker

To understand earth science, you must combine the methods and evidences of both science and history. If you also use the "history book of the world," the Bible, you can make sense of the earth's surface.

P 978-0-89051-800-7 $15.99 U.S.

Building Blocks in Science
Gary Parker

Explore some of the most interesting areas of science: fossils, the errors of evolution, the evidence of creation, all about early man and human origins, dinosaurs, and even "races." High school.

P 978-0-89051-511-2 $15.99 U.S.

Building Blocks in Life Science
Gary Parker

Teaching high school level science from a biblical perspective. Teachers and students will find clear biological answers proving science and Scripture fit together to honor the Creator.

P 978-0-89051-589-1 $15.99 U.S.

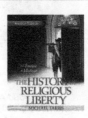

The History of Religious Liberty Student Edition
Michael Farris

This special edition was designed to be used with the high school course *The History of Religious Liberty.* This student-friendly text has been enhanced with images of relevant people, places, and events!

STUDENT: P 978-0-89051-882-3 $34.99 U.S.
TEACHER: P 978-0-89051-869-4 $15.99 U.S.

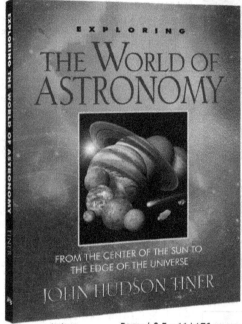

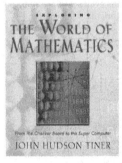

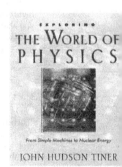